CRISIS IN THE ATMOSPHERE

THE GREENHOUSE FACTOR

ED PHILLIPS

D.B. CLARK & CO. PUBLISHING

PHOENIX, AZ USA

CRISIS IN THE ATMOSPHERE: The Greenhouse Factor

ISBN 0-9625245-0-6

Art Direction & Design: Robin Fox & Associates
Production Services: The Art Studio

0 9 8 7 6 5 4 3 2 1
F R C A M C P D S M C P

Library of Congress Catalog Number: 90-93256

For information contact the Publisher:

D. B. Clark & Co.
International Division
P.O. Box 35022
Phoenix, AZ 85069

PUBLISHED IN THE UNITED STATES OF AMERICA
PRINTED ON RECYCLED PAPER

About The Book

The six hottest years on record have all occurred in the 1980s. Most scientists attribute the hot temperatures to the increased greenhouse effect. During the 1980s scientists also discovered a hole in our ozone layer over Antarctica that is over 2,000 miles in diameter — roughly the size of the United States.

Crisis in the Atmosphere: The Greenhouse Factor reports the facts from man's abuse on our atmosphere. It explains, in simple terms, how we are warming our atmosphere with the increased greenhouse effect and how we are destroying the ozone layer. The book also discusses the consequences we face from warmer temperatures and less ozone. An easy-to-understand chapter explains what we can do to help correct these environmental problems.

Crisis in the Atmosphere: The Greenhouse Factor is an informative and easy-to-understand book on key environmental problems facing us today.

CRISIS IN THE ATMOSPHERE

THE GREENHOUSE FACTOR

D. B. CLARK & CO. PUBLISHING

PHOENIX, AZ USA

To Anne and Alex

Contents

Chapter 1

THE GREENHOUSE FACTOR: Fact or Fiction?

The six hottest years on record have all occurred in the 1980's.

Global warming is the result of the continued release of heat-trapping pollutants into our atmosphere. The continued abuse of our atmosphere since the Industrial Revolution is starting to take its toll. In 1982, scientists discovered a seasonal ozone hole over Antarctica that's roughly the size of the United States. Now there is strong evidence of ozone depletion over the Arctic. Climatologists are predicting that global temperatures could rise by as much as 9 degrees Fahrenheit (°F) during the next 50 years.

As man releases more and more pollutants into the atmosphere, the atmosphere fights a battle to maintain an equilibrium. Until now our atmosphere has been doing a good job of taking care of both the man-made and the natural pollutants. It is the man-made pollutants that are now forcing our atmosphere out of balance.

We force our atmosphere to work overtime to fight a daily influx of gases created by industrial activities. The burning of fossil fuels is one of the chief sources of greenhouse gases. Fossil fuels emit carbon dioxide, which is very efficient at trapping heat. These gases trap heat and cause the earth's temperature to rise. Fossil fuels, such as coal, oil and natural gas, account for about half of the greenhouse effect. Since the mid-1800's, carbon dioxide levels in the atmosphere are up by 30%. The latest survey by the United States Environmental Protection Agency reveals that other gases are entering our atmosphere. Here is a sample of annual releases in the United States:

- 55.7 million tons of carbon monoxide
- 19.6 million tons of ozone-producing compounds
- 18.5 million tons of sulfur dioxide
- 17.7 million tons of nitrogen dioxide
- 7.3 million tons of lead
- 1.3 million tons of 308 various toxic chemicals

FACT OR FICTION ?

The world is clearly warmer now than it was a century ago. After 150 years of significant atmospheric pollution, the average global temperature has risen 1° F in the past 100 years. It may not sound like much, but considering it took 2,000 years for the earth to warm 1° F before this, that's a significant increase. It now appears that the increased greenhouse effect is playing a major role in setting record temperatures throughout the world.

It is a fact that ozone levels over Antarctica have declined dramatically as a result of the release of man-made ozone-eating gases into the atmosphere. In October, 1989 ozone levels registered a decline of more than 50% in the atmosphere over Antarctica. New research has shown the ozone over the Arctic is also decreasing.

THE GREENHOUSE FACTOR

The unintentional chemistry experiment taking place in our atmosphere continues. Scientists are at a loss to estimate when nature will say enough is enough. Eventually, there will be a point at which the earth's atmosphere will severely react to all the pollutants and gases we put into the atmosphere.

> **The GREENHOUSE FACTOR is the critical threshold at which severe and unpredictable reactions will take place in the earth's atmosphere.**

We know that water boils at 212° F and freezes at 32° F. There are certain thresholds at which chemical and physical reactions take place. What is the threshold or breaking point of our atmosphere? When will our atmosphere "breakdown" because of continued abuse? It is not possible to model our atmosphere accurately on a computer — it is too complex with too many unknowns. Calculating what will happen is difficult. We do know, however, that climate changes are occurring and that we are experiencing record temperatures. Some atmospheric breakdowns are beginning to appear. Scientists have discovered an ozone hole in our atmosphere. With significant ozone depletion over the Arctic (North Pole), and the entire Northern Hemisphere, global ozone depletion is approaching 3%.

The Greenhouse Factor is approaching reality. We are starting to upset the natural balance of our atmosphere.

THE GREENHOUSE EFFECT

The cause of the greenhouse effect is no mystery. Certain gases float upwards into the air and remain there. These gases, called greenhouse gases, partially block and trap the escape of heat back into space. Left on its own, the earth has managed to keep an equilibrium in the atmosphere. In fact, the earth maintains its own natural greenhouse effect. Without greenhouse gases, our planet would be about 72° F cooler. Our current average global temperature is about 59° F. This natural greenhouse effect provides a warm blanket of air around our planet and makes life possible. Disturbing the delicate balance of the earth's atmosphere by the release of polluting, heat-trapping gases results in higher temperatures. More gases in our atmosphere trap more heat. Warming the air causes a more pronounced greenhouse effect.

The United Kingdom Meteorological Office has been collecting world temperature data for more than a century. Their data shows that the hottest year on record is 1988, and the second hottest year is 1987. The six hottest years of the entire 20th century have all occurred since 1980. Beginning with the warmest years they are: 1988, 1987, 1983, 1981, 1980 and 1986.

Throughout America, record high temperatures in the 1980's were a reality. During that decade, climatologists recorded more record high temperatures since regular and comparable temperature recordings began. (See the list of temperature records in the Appendix.)

OUR OZONE LAYER AT RISK

Record high temperatures are not the only result of the abuse our atmosphere is taking. We are also destroying our ozone layer. Our atmosphere constantly manufactures ozone through a chemical reaction with sunlight. Ozone destruction occurs naturally in the atmosphere, too, balancing this production. However, man is releasing certain chemicals into our atmosphere disrupting this balance. Man is destroying ozone faster than our atmosphere produces it. The result is a thinning, and even the disappearance, of parts of our ozone layer.

Ozone is the shield that protects life from the harmful rays of the sun. Our ozone layer filters out the harmful ultraviolet or UV rays. Our skin would burn almost instantly without the ozone above. To travel outside we would have to cover up completely to avoid serious sunburn. UV blocking sunglasses would be essential. Less ozone above us would cause significant lifestyle changes.

Unfortunately, we can't shade our food crops from the harmful UV rays. Too much ultraviolet radiation damages sensitive tissues and even the genetic structure in plants. Food production would decrease dramatically with less ozone. This would, in turn, threaten the already-scarce food resources in many parts of the world. Famine would result in the poorer countries, and a dramatic lowering of the standard of living would result in the wealthier countries.

THE CHEMISTRY EXPERIMENT ON OUR ATMOSPHERE CONTINUES

We don't know what the tolerance level is in our atmosphere. At what level do we start causing

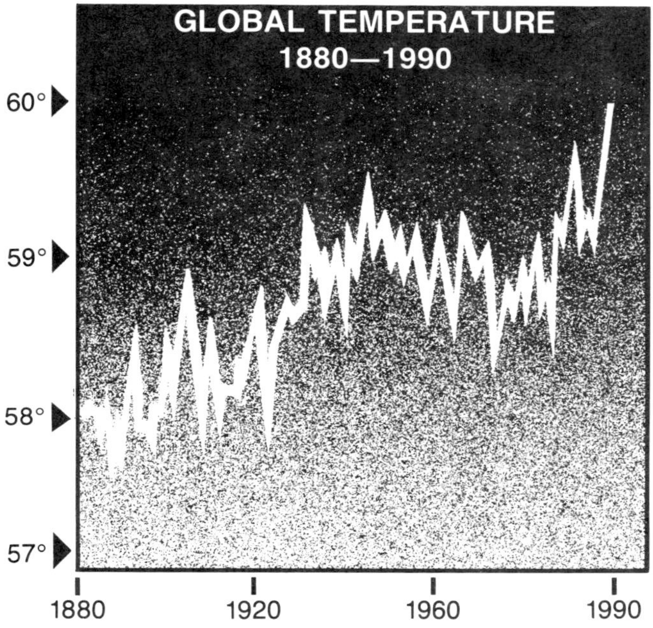

The average global temperature has risen 1°F in the past 100 years.

permanent damage? There will be recurring high temperatures and stronger and more frequent tropical storms and hurricanes if we keep disturbing the balance in the atmosphere. No scientist or climatologist really knows what all the chemical reactions above us will cause. Scientists *discovered* the ozone hole. They did not *predict* it. Destruction of ozone is science-fact, not science-fiction. It strongly appears that more surprising and even frightening discoveries are inevitable.

We are crossing the threshold of atmospheric abuse — The Greenhouse Factor.

Chapter 2

THE CREATION OF THE GREENHOUSE EFFECT

We have our own natural greenhouse effect here on earth. Without the greenhouse effect life would certainly be difficult.

As a man-made greenhouse lets the sun pass through the glass but traps the heat inside, the greenhouse effect warms our planet. The greenhouse effect makes our planet a comfortable place in which to live. Think of the last time you left your car in the sun with all the windows closed. On a cold day, the air inside your car is warmer than the air surrounding it. On an already hot day, the inside of your car gets unbearably hot. For most of us, getting burned on the car steering wheel is our first painful example of how the greenhouse effect works. A similar occurrence is taking place in our atmosphere. The polluting gases released into the atmosphere are trapping the reflected heat from earth. It is a lot like the car windows trapping the warm air. Scientists call these heat-trapping gases greenhouse gases.

Our atmosphere consists of about three-fourths nitrogen and a little less than one-fourth oxygen. These two gases have little to do with creating the greenhouse effect. The remaining fraction, about 1%, makes up the "trace gases." These trace gases cause virtually all of the greenhouse effect. Some of these trace gases are greenhouse gases. The most significant single greenhouse gas is carbon dioxide, followed by methane, CFC's (chlorofluorocarbons) and nitrous oxide.

THE GREENHOUSE EFFECT

The greenhouse effect is a widely known principle in atmospheric science. Solar radiation (sunlight) penetrates the atmosphere and warms the earth. The earth then re-radiates this heat energy back into the atmosphere. Some of the heat escapes into space. This reflected heat is a different form of energy and some of this heat remains trapped in the atmosphere. The trace or greenhouse gases trap the reflected heat energy from earth. This creates the greenhouse effect.

The sun's energy coming to earth is in "short wavelengths." This shortwave energy easily penetrates the atmosphere and reaches the earth. When it hits the earth, the earth absorbs some of its energy and reflects the rest. The reflected heat energy is "long wavelength" or infrared energy. The greenhouse gases then trap this reflected or infrared energy. More greenhouse gases in the atmosphere serve to trap more heat. This changing of the natural chemistry of the air by emitting more greenhouse gases is what is creating the increased and pronounced greenhouse effect.

GREENHOUSE GASES

It is several of these trace gases that are causing our greenhouse problem. Greenhouse gases are necessary to our survival since these gases provide our planet's natural greenhouse effect. However, we now face a problem of an overabundance of greenhouse gases. The key culprits here are: carbon dioxide, methane, CFC's, and nitrous oxide.

CARBON DIOXIDE

As early as 1861 scientists discussed the idea of temperature changes in climate due to rising levels of carbon dioxide. Since 1896, some scientists have voiced a concern about the effects of too much carbon dioxide in our atmosphere. In that year, Swedish scientist Svante Arrhenius published a paper that described the likely effects on climate due to the release of carbon dioxide from burning coal. In 1899, American scientist P. C. Chamberlain also wrote about the effects on our climate from too much carbon dioxide in the atmosphere. Today, nearly a century later, articles from scientists all over the world are still discussing, and often warning of, the problem of too much carbon dioxide in the air.

Carbon dioxide is a colorless, odorless, and tasteless gas. It is the most important greenhouse gas and represents about half of the greenhouse problem. On the other hand, carbon dioxide produces many benefits. It helps plants thrive. It is frozen to make "dry ice," and it even makes beer and soft drinks fizz.

Carbon dioxide (CO_2) is a by-product from burning fossil fuels such as wood, coal, petroleum products, natural gas and other resources containing carbon. Thus, the burning of the tropical rain forests significantly adds to the production of CO_2. Trees and other plants breathe in and store CO_2 and release oxygen.

The impact of CO_2 upon the atmosphere appears insignificant. It makes up less than one-half of 1% of the total atmosphere. Yet despite the low concentration, CO_2 plays an important role in determining our planet's temperatures. CO_2 is very efficient at absorbing the heat reflected off our planet.

The amount of carbon dioxide in the atmosphere of our planet has been increasing at an alarming rate since the mid to late 1800's. The amount of CO_2 has increased nearly 30% since 1860. In the mid-1800's, CO_2 levels were 270-280

The Creation of the Greenhouse Effect

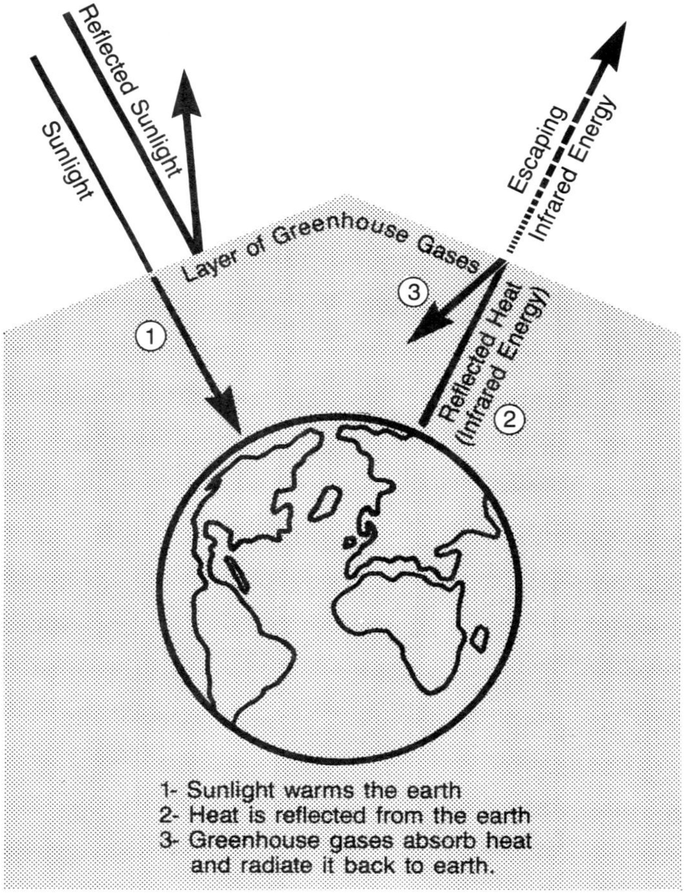

An increase in the concentration of certain gases in our atmosphere causes more heat to be retained, resulting in higher temperatures.

parts per million parts of air (PPM). In 1957, the reading was 315 PPM. (See graph.) In 1990, CO_2 levels stood at 350 PPM.

CO_2 readings are also very reliable, since key monitoring stations are away from major cities and industrial areas. One very important CO_2 monitoring location is on the top of Mauna Loa 11,000 feet high on the Island of Hawaii. Daily observations of the CO_2 content in the air occur year-round. The site was chosen because it sits in an area of solid black lava. This area provides undistorted readings from CO_2 uptake and release from green plants and trees during photosynthesis and respiration. Most of the increase of CO_2 has occurred since 1960. Many scientists predict that CO_2 will reach 1,000 PPM —more than doubling current levels — by the year 2050. The increase is significant and the current rate of increase is alarming.

Left alone, the supply of CO_2 has maintained a balance for most of earth's existence. Then came the Industrial Revolution and the rampant burning of wood and coal. The timing of the increase in CO_2 in the air parallels the Industrial Revolution that started in the mid to late 1800's. Factories burned fuels which produced a large amount of CO_2. Today, factories still produce record amounts of CO_2. Combustion of gasoline in cars also generates carbon dioxide.

The Rise in CO_2

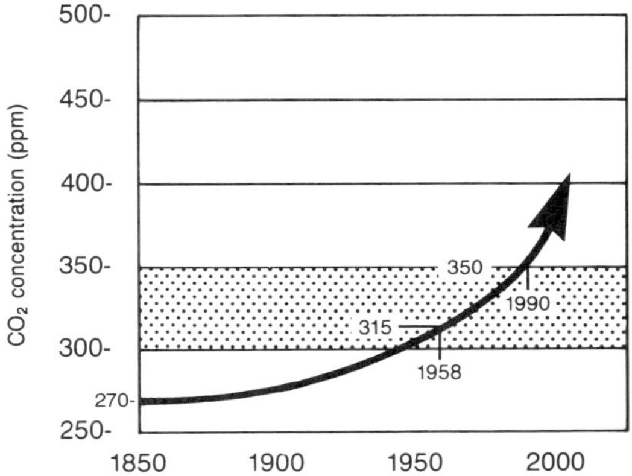

Carbon dioxide levels in our atmosphere have risen 30% since the mid-1800s.

The current estimate is that man produces over 20 billion tons of carbon dioxide each year from fossil fuels. The United States accounts for about 20% or 4 billion tons of these carbon dioxide releases. In the United States, with our energy consumption, we produce an estimated average of nearly 15 tons of CO_2 each year for every person. The burning of one gallon of gasoline puts over 20 pounds of CO_2 into the atmosphere. Nearly 60% of global warming is a direct result of fossil fuel related energy production.

Fortunately, of the over 20 billion tons of CO_2 produced every year, less than half actually goes into the atmosphere. About 60% of the current CO_2 production is taken in by plants or stored in the oceans. The remaining 40% goes into our atmosphere.

Do we know that the amount of CO_2 is increasing over time? One way to find out is to test a sample of air from the past. When snow turns to ice, it traps air at about 10% of its volume. The ice readings from glacial areas are a very valuable resource for finding out what our air looked like thousands of years ago. Scientists from the University of Bern, Switzerland tested bubbles of air found in ice that formed in the Antarctic and Greenland 1000 years ago. They found that this air contained approximately 270 PPM of CO_2, about the same found in the 1800's. The method is valid to within 10 parts per million.

As further proof and analysis, research of tree rings suggests that the amount of CO_2 changed little from 1000 years ago to just before the Industrial Revolution. The research showed the amount of CO_2 in the air to be about 260 to 270 PPM.

As we move closer to the present, it becomes more alarming. All the tests of air trapped in polar ice and tree ring analysis show the rate of CO_2 accelerating. The acceleration corresponded to the rise in industry beginning in the 1800's.

During the International Geophysical Year in 1957-1958, two sites were chosen to monitor CO_2.

The South Pole and the top of Mauna Loa in Hawaii were chosen because they were far from industrial pollution sources. By 1957, measurements from these stations showed the amount of CO_2 in the air had increased to 315 PPM. That is a 17% increase from 270 PPM in less than 100 years. This is certainly a significant increase, but perhaps not alarming. Current measurements of the amount of CO_2 in the air are at 350 PPM. That is an increase of about 30% since before the Industrial Revolution. About a third of the CO_2 increase in our atmosphere has occurred in just the last two decades. Although CO_2 is up, it is now the *rate* of increase in CO_2 that is sounding the alarm on the greenhouse effect. Further, ice samples now reveal that during the past 160,000 years — covering two ice ages — CO_2 levels have never exceeded 270-280 PPM. In all of recorded history, never has the global CO_2 level been as high as it is now at 350 PPM.

RISING CARBON DIOXIDE LEVELS TRAP HEAT

Why is increasing CO_2 a concern? More CO_2 in the air traps more heat in the lower levels of the atmosphere. That means a warmer climate. The atmosphere absorbs little of the sun's radiation as it passes through. Much of the sun's radiation warms the earth's surface. The reflected or re-radiated heat from earth then encounters carbon dioxide and other greenhouse gases in the atmosphere. Some reflected heat escapes back into space. As the amount of CO_2 increases in the atmosphere, less heat escapes back into space. More heat stays in the air close to our planet and that means a warmer climate. A warmer climate is one of the results of the increased greenhouse effect.

OUR ATMOSPHERE AND CARBON DIOXIDE

This warmer climate, however, is what allows us to live on our planet. Without our atmosphere and the greenhouse effect, the earth would just be too cold to support life as we know it. Without our atmosphere, the average temperature near the surface of the earth would be about -13° F. That compares to the current average temperature near the surface of 59° F. The natural greenhouse effect is necessary. It keeps us out of the deep freeze.

Without our blanket of atmosphere to warm temperatures here on earth, living on the earth would be more like living on the moon. Without an atmosphere, the climate on the moon is brutal. Unchecked sunshine bombards the moon with ultraviolet rays and no life exists there. Water would boil in the heat of the day as the temperature at the surface soared to 260° F. By contrast, at night, with no air to help hold in the heat, temperatures plummet to 240° F *below* zero. Our nearest neighbor in the solar system, the airless, waterless moon is even less habitable than the Antarctic or the Sahara Desert.

Mars is a little more than half the size of the earth but closer to the moon in climate. The atmosphere is rich in CO_2 yet there is just too little atmosphere to cause a significant greenhouse effect. Temperatures on Mars rise during the day to 68° F above zero. Without the blanket of atmosphere and greenhouse effect, temperatures plunge to 300° F below zero at night.

Venus is the nearest planet to earth, and nearly the same size. It is an example of the greenhouse effect out of control. The atmosphere is very dense on Venus and rich with CO_2. In fact, it consists almost entirely of carbon dioxide. These are perfect ingredients for a greenhouse effect. The radiation from the sun easily penetrates the atmosphere and the clouds that shroud Venus. The CO_2, clouds and thick atmosphere trap the reflected heat. The CO_2-rich atmosphere traps so much heat on Venus that the surface must be like what we imagine in Hell. Storm clouds boil and churn and blast the planet with lightning. The clouds are so thick that it is dark and foggy. Temperatures can reach 900° F — hot enough to melt lead. Venus is an example of a runaway greenhouse effect — a greenhouse effect out of control.

Compared to our nearest neighbors, the earth is clearly paradise, thanks to our built-in or natural greenhouse effect.

Although CO_2 is the major cause of the greenhouse effect, there are other greenhouse gases.

METHANE

Methane is the second largest contributor to the greenhouse effect. This greenhouse gas causes about 18% of our greenhouse problem. Methane is another of the trace gases in our atmosphere. It is 25 times more efficient than CO_2 at absorbing the reflected heat (infrared radiation) from the earth. Methane is produced when oxygen is not freely available, such as in swamps, bogs, rice paddies and moist soils. In recent years, there has been about a 1% rise per year in methane. As was the case with carbon dioxide, scientists used ice samples to determine previous atmospheric levels of methane. Like the levels of carbon dioxide, methane levels also revealed significant increases from the mid-1800's to the present. In 1951 methane registered 1.1 PPM and today the levels are more than 1.7 PPM. Methane levels, like carbon dioxide levels, have increased significantly in our atmosphere as a direct result of human activity. About 60% of total methane releases is human-directed while the other 40% comes from natural methane sources such as swamps, marshes and the oceans. Since methane is much more efficient at trapping heat from the earth than carbon dioxide, there is concern over the ever-growing releases of methane. Like carbon dioxde, methane levels are at record highs. They were fairly stable in the last 10,000 years according to ice-core samples. In the 1800's, the methane levels took off and have now more than doubled.

Methane is the major constituent of natural gas. Methane also results from forest fires, digestive tracts of animals, rice paddies, coal mines, landfills and termite mounds.

There is a growing concern of the rise in methane from livestock. The increasing food needs of the world's population demand more domesticated livestock. Domesticated livestock number more than 3 billion, and of that, cows number over 1.2 billion. Cows can belch about 100 gallons of methane per day, and in total, release about 100 million tons of methane annually. Methane gas released from livestock accounts for about 15% of the total methane gas build-up in our atmosphere.

Rice paddies are another major contributor of methane to our atmosphere and account for about 20% of the methane problem.

Oil and natural gas releases add another 8% to the problem while coal mining contributes 6%.

There is one additional large potential source of future releases of methane that now concerns scientists. This methane is "locked up" in permafrost and in frozen tundra areas. Rises in global temperatures may cause the permafrost to melt and tundra to thaw. This would cause large releases of methane into our atmosphere. Our atmosphere would then have to contend with a very efficient heat-trapping gas. The rise in methane would cause significant atmospheric warming. We do not know what these added releases would do to our atmosphere. Again, we would approach the Greenhouse Factor because we would enter new thresholds — the likes of which we have never experienced.

CFC's

Another greenhouse gas, and perhaps the most widely known, are chlorofluorocarbons or CFC's. These completely man-made gases cause about 14% of the greenhouse problem. CFC's are about 10 times more efficient than CO_2 at trapping heat.

There are many types of CFC's. Some are known under the trade names of Freon and Styrofoam. We use CFC's to cool refrigerators and we use them in air conditioners. Some food containers, such as those that keep hamburgers warm and drinks warm or cold, contain CFC's. CFC's also propel aerosols in such products as deodorants and hairsprays.

The United States banned the use of CFC's in most aerosol sprays in the late 1970's. Unfortunately, the ban covers only a fraction of total CFC uses in the United States. Invented in 1928, CFC's are relatively new to our atmosphere. CFC's not only serve as a greenhouse gas in trapping heat in the atmosphere, but they also filter upwards to the stratosphere, where they eventually destroy ozone.

OTHER TRACE GASES

Other trace gases which add to the greenhouse effect are nitrous oxide, ground-level ozone and water vapor.

Nitrous oxide, or "laughing gas," adds about 6% to our greenhouse problem. Nitrous oxide, or N_2O, has also exhibited significant increases. N_2O has gone from a level before the Industrial Revolution of 280 parts per billion to a current level of 350 parts per billion. Each year our atmosphere receives about 5 million tons of nitrous oxide. The bad news is that N_2O is very stable and can last upwards of 150 years in the atmosphere. It is also much more efficient than CO_2 in trapping heat. One major source of rising levels of nitrous oxide is from the artificial nitrogen fertilizers used in agriculture. The other source is the burning of fuels that contain N_2O such as coal. Nitrous oxide also helps destroy the earth's ozone layer.

Ground-level ozone is also a greenhouse gas. Sunlight reacts with car exhausts to form this polluting and poisonous gas. When ground-level ozone rises, it acts as a greenhouse gas and traps heat.

Water vapor is a naturally occurring greenhouse gas. It does a good job at trapping reflected heat. Ironically, man is creating more water vapor in the atmosphere since rising global temperatures have been increasing evaporation from surface water.

None of the greenhouse gases except CO_2 individually pose a short-term serious threat to our atmosphere. It is the combined total of these gases that threaten to increase our greenhouse effect. These other "trace gases" taken together are significant. They affect the greenhouse as much as CO_2. That means all the other trace gases will probably cause about half of the future greenhouse effect.

DEFORESTATION

In addition to the increase in greenhouse gases, deforestation (the cutting down of trees) is also adding to the greenhouse effect in a big way. About 80% of the total annual discharge of CO_2 into the atmosphere results from the combustion of coal, oil and gas. Twenty percent of the CO_2 problem comes from deforestation — mostly in the Tropics. Every day, about 100 square miles of rain forest go up in smoke.

Trees, as well as most other plants, take in CO_2 and convert it to oxygen when sunlight is present. This process is called photosynthesis. In addition to this conversion process, plants and trees store CO_2. The cutting down or burning of trees

Sources of Greenhouse Gases

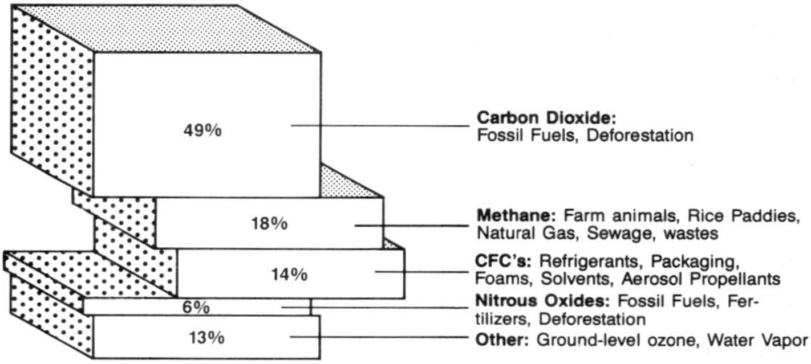

Certain gases, known as greenhouse gases, trap heat and cause temperatures to rise.

releases the stored CO_2, which then acts to trap reflected heat from the earth. Dead trees can no longer store CO_2 or convert it into oxygen. Thus, more and more CO_2 collects in the atmosphere.

OUR ATMOSPHERE IS ONLY SO LARGE

Our immediate atmosphere, the troposphere, is only about 5 miles high at the North and South Poles and only 10 miles high at the equator. The troposphere contains about 85% of our atmosphere. Above the troposphere, the stratosphere contains the remaining 15%. Our atmosphere is not infinite as most people think. It is not very large when you consider that the typical distance that Americans travel to work is around 10 miles. If you traveled to work straight up you would be above 85% of the atmosphere and would need oxygen to breathe. There is only so much room for chemicals, gases and other pollutants to occupy before a breakdown occurs. We are threatening the natural balance of our atmosphere. We are moving into uncharted territory with the immense releases of man-made gases.

The problem is: we are just putting too many greenhouse gases into the air. Some of these gases take years to migrate upward and do their damage. The extra gases upset the delicate balance of natural chemical reactions. This in turn is leading to the pronounced greenhouse effect and the destruction of the ozone layer. Are we too late to stop it? The jury is still out. There is a threshold where our atmosphere cannot maintain a natural balance. Breakdowns are beginning to appear. Warmer global temperatures, changing climate patterns, and the appearance of a continent-sized ozone hole argue that the Greenhouse Factor is here.

Chapter 3

OZONE: The Shield That Protects Us

There are two major types of ozone. The most widely known is the layer of gas around the earth that protects us from the sun's harmful rays. We call this the ozone layer. The second is man-made or man-caused ozone found at ground level. Let's look first at the atmospheric ozone or our ozone layer.

ATMOSPHERIC OZONE: THE OZONE LAYER

Ozone resides in the stratosphere, about 12 to 30 miles up. The highest concentration of ozone is in the area of 12 to 20 miles. It is the shield that protects us. Our atmosphere constantly produces and destroys ozone. Left alone, ozone maintains a balance in the stratosphere. This balance provides a superb blanket for protecting us from the ultraviolet radiation or harmful rays of the sun. Ozone's enemy is man. Certain pollutants that man releases into the atmosphere diminish ozone production.

Ozone is sparsely distributed throughout the stratosphere. It is so diffuse that if it were compressed and brought to the surface, it would only cover the ground a hundredth of an inch thick. A good analogy is to compare the outer skin of an apple to our ozone layer—it is very, very thin.

It is much warmer in the stratosphere versus the upper portion of the troposphere. This is because ozone in the stratosphere absorbs a portion of the incoming sun's rays (UV rays) and converts them to heat. The ozone is broken down into atoms of oxygen. The energy in this chemical process is what actually heats the stratosphere. These atoms of oxygen then recombine into ozone and the process continues over and over.

With a decreased ozone layer, more UV rays from the sun reach earth. The consequences of an increase in UV rays reaching earth are significant: more skin cancer, crop destruction, changing climate, alterations in the food chain, and even effects on our human immune system.

Ozone Location

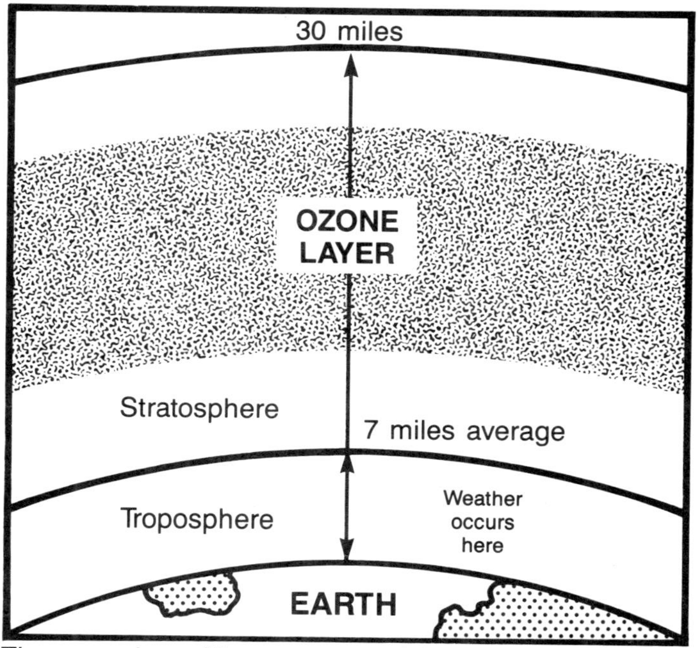

The ozone layer filters out most of the harmful UV rays of the sun. Global ozone levels have fallen nearly 3%.

GROUND-LEVEL OZONE
(Man-Caused Ozone)

Ozone found at ground level is actually a family of chemicals called photochemical oxidants. Ozone is the main part of smog. To produce ground-level ozone requires the right chemicals plus sunlight. When autos and other sources put hydrocarbons and various organic gases into the air these gases rarely stay in their original form. Sunlight shining on these gases in the presence of nitrogen oxides (usually from autos) causes complex chemical reactions that form ground-level ozone. For this reason, ground level ozone is highest in daylight hours. Since ozone is heavier than air, it tends to stay near the ground where it forms. This makes for high concentrations of ozone in and around major freeways and busy streets.

Other sources of ground-level ozone besides fumes from gasoline include: volatile liquids such as dry-cleaning fluids, hazardous wastes and even smoke from forest fires. Ozone gives off a pungent odor that is most noticeable such as when lightning occurs during a storm. It is also the smell you notice around electric train sets.

Ground-level ozone creates problems. It is deadly in large doses and is very harmful to plants and animals even in small amounts. That is why newscasts and weather reports often mention ozone levels. When humans or animals breathe in ozone, inflammation and a stiffening of the lungs results. When ozone levels are high, people should reduce outdoor activity. Only 10% of the ozone we inhale is exhaled. The other 90% remains in our bodies. Since ozone levels are highest during daylight (sunlight reacts with car exhausts to produce ozone), joggers should avoid heavily traveled streets.

The United States Environmental Protection Agency (EPA) has set .120 parts per million (PPM) of ozone in the air as the maximum safe limit. On a daily basis, some of America's largest cities and metropolitan areas exceed this safe limit. Most of the entire eastern seaboard regularly surpasses safe ozone limits as does southern California. Large metro areas are not the only areas that exceed safe ozone limits. In 1988, the rural state of Maine and areas in northern New York State (near Canada) both recorded unsafe ozone readings. Before 1988, there were no excess ozone levels in these two rural areas.

The average ozone level at ground-level in the U.S. was 5% higher in 1988 than it was in 1983. As with the case of the greenhouse gases discussed in Chapter 2, ground-level ozone is at record highs and continues to build daily.

Ground-level ozone is also a greenhouse gas. This further compounds the problem as we now have to contend with this gas as it migrates upward. Also, warmer temperatures brought about by the greenhouse effect result in significant increases in ground-level ozone.

At ground level, concentrations of ozone can cause increased incidence of emphysema and other pulmonary diseases. Ozone irritates the eyes and in higher concentrations, reduces stamina in humans.

Human interference is knocking our atmosphere out of its natural balance. CFC's and the burning of fossil fuels are destroying record amounts of ozone. This new type of chemical warfare on our atmosphere is escalating. We do not know when our atmosphere will reach the critical threshold where severe and unpredictable reactions will take place. Total ozone has declined by almost 3% in the last 20 years. Pollution of our skies has caused this decline.

What are these pollutants that destroy our ozone layer? Here are the main culprits:

CFC's

CFC's or chlorofluorocarbons are not naturally present in our atmosphere. Man produces CFC's. We know some of them by the trade names of Freon and Styrofoam. Once these CFC's enter the atmosphere, they float upwards 12 or more miles until they encounter the ozone in the stratosphere. Under certain conditions, the CFC's then destroy ozone. Sunlight changes the molecules' structures, creating new ones that merge with and then destroy the ozone that protects us.

CFC's were industry's golden boy when developed in 1928. They are odorless, nontoxic, nonflammable and noncorrosive. They are the wonder chemicals that keep our refrigerators and air conditioners cool. More than 90 million cars in the United States are air conditioned and use CFC's. Many aerosol cans also contain CFC's as a propellant for certain products. Many disposable food containers like the styrofoam containers that keep hamburgers warm, contain CFC's. An even greater threat to the ozone layer is the large amount of CFC gases that escape during the manufacturing process of these containers.

It was not until 1974 that scientists found a link between CFC's and ozone destruction. The chlorine in the CFC's acts to destroy ozone. In fact, **one** single CFC-chlorine molecule, during its lifetime, can destroy as many as **100,000** ozone molecules. Not only do CFC's cause ozone destruction, CFC's are also a greenhouse gas. They are very effective in trapping reflected heat from earth. Thus, CFC's are doing double duty in the atmosphere.

As the CFC's destroy ozone in the upper stratosphere, more ultraviolet rays pass down to the lower stratosphere where the suns rays may interact with oxygen to produce ozone. The problem is that now ozone is much closer to earth. As a greenhouse gas, it's a good absorber of the reflected infrared or reflected sun rays from earth. The ozone thus traps reflected heat from earth and further enhances the greenhouse effect.

During the mid-1970's, production reached more than two billion aerosol cans annually. In 1978, the United States banned CFC's in most aerosols. Unfortunately, the ban did not cover all aerosol uses. The ban covered less than 25% of global use of CFC's. In spite of the ban, the U.S. remains the world's top producer of CFC's.

According to statistics, world industries produce about 2.5 billion pounds of ozone-eating CFC's every year. The U.S. accounts for a little less than one-third of the total. Production of CFC's continues to increase at 4% annually.

Another serious ramification of CFC's is that they are very stable. Once released, they can remain in the atmosphere for up to 100 years before they break down. If we were to stop making all CFC's today, the problem would still be with us for 100 years.

NITROGEN OXIDES

Chemists call nitrogen oxides NO_x. The burning of jet fuel by commercial aircraft releases large amounts of NO_x into the lower stratosphere and upper troposphere. These releases of NO_x about 11 miles above ground can result in significant reductions in atmospheric ozone (the ozone layer). This occurs indirectly when the different nitrogen oxides combine with ozone, resulting in the ozone changing to oxygen and other oxides of nitrogen. The Concorde and military supersonic aircraft (SST's) fly at even higher levels in the stratosphere. This puts the NO_x emissions closer to the ozone layer. Obviously, the pollution from high flying aircraft is even more destructive to our ozone layer.

Nitrous oxide is the scientific name for laughing gas. Unfortunately, it is no laughing matter when released into the atmosphere. Every time we burn coal or oil, this high temperature combustion produces some nitrous oxide. As the nitrous oxide rises into the stratosphere it contributes to ozone depletion.

NATURAL OZONE EATERS

In addition to contending with CFC's and nitrous oxides, the ozone layer fights a daily battle with natural ozone eaters to maintain a balance. Certain natural events on earth contribute to ozone destruction. These include: volcanoes, changes in the earth's magnetic field, and extraterrestrial events like sunspots and cosmic rays. Nature does a good job of keeping natural forces in check. It may be asking too much to expect nature continually to work overtime against both nature's and man's releases.

LESS OZONE MEANS MORE ULTRAVIOLET LIGHT

Life on earth is dependent on solar radiation as its source of energy. All living organisms require energy for growth, development and reproduction. The process of photosynthesis in green plants converts sunlight, carbon dioxide, and water into carbohydrates rich in energy. Organisms that cannot produce their own energy through photosynthesis must participate in a food chain that includes plants that can.

We need solar radiation to sustain life as we

know it. Up until this decade we had all we needed. Now, because of mismanaged atmospheric pollution causing a thinning of our ozone layer, we are getting too much. We know now the biological results of too much solar radiation (UV rays), especially the UV-B rays.

UV-B RAYS

Less ozone up above means that more harmful ultraviolet radiation, or UV rays, will reach us. The most severe UV rays are those of the shorter wavelengths of the ultraviolet spectrum called UV-B rays. Human skin easily absorbs UV-B rays. While UV-B rays are helpful in the human production of Vitamin D, too many UV-B rays are harmful. That is why many sunglasses contain UV-B filters in the lenses and why people use sunblock while in the sun.

Studies clearly prove that severe consequences follow above-average exposure to UV-B radiation. The UV-B rays can cause innumerable problems, some of which follow:

SKIN CANCERS

Doctors diagnose more than 500,000 cases of skin cancer each year in the United States alone. If ozone depletion continues, skin cancer will rise significantly. Each 1% drop in ozone allows 2% more of the UV-B rays to reach the ground, increasing skin cancer cases from 3-6%. That could mean almost 50,000 more skin cancer cases a year in the United States. UV-B, the shorter wavelengths of the ultraviolet spectrum, are the cause of virtually all skin cancers. The damage caused by UV-B to the skin cells is cumulative over years of exposure. As the ozone layer thins, light-skinned people are most at risk. More UV-B reaching earth in small doses may be too subtle for most to notice. As ozone decreases, sunburns will get worse and become more frequent. For every 1% drop in ozone, an additional 40,000 people will die from skin cancer during the next two generations. Fortunately, most skin cancers are treatable, and if detected early, are curable. About 8,000 Americans die from skin cancer each year. Unfortunately, malignant melanoma cancers have increased more than 100% in the 1980's. The medical community predicts all cases of skin cancer will double again in the next 25 years.

HUMAN IMMUNE SYSTEM

UV-B also damages the human immune system. Exposure to increased doses of ultraviolet light causes immunological changes. It impairs cells that fight infections and skin cancers. UV-B actually triggers the production of cells that suppress certain human defense mechanisms. Recent evidence shows that sun screens may not stop the harmful immunological effects of UV rays. Ultraviolet light also aggravates certain diseases such as herpes. We do not know other potential effects on the human immune system from UV-B rays. There are many unknowns as to what havoc may be done to the human immune system.

CATARACTS

Ultraviolet radiation is a major cause of cataracts, a clouding of the lens of the eye. Untreated, cataracts cause blurred vision and eventually blindness. The incidence of cataracts rises with increased exposure to UV-B rays.

CROPS

Perhaps the hardest hit victims of ozone depletion will be our crops. They are very vulnerable to UV-B rays. The rays cause cell and tissue damage in many crops. A reduction in our ozone layer will result in a corresponding destruction of crops. A one second exposure to ultraviolet radiation with no ozone present is lethal to living cells. Increased incidence of ultraviolet light may have a material effect on agriculture. A depletion in the ozone layer causes the genetic material of plants and crops to absorb more ultraviolet light, thus changing their basic genetic foundation. For species to continue, the genetic material must be very stable. UV light causes mutations. Since we know very little on this subject, the decreasing ozone layer may have immense and unforeseen implications.

OCEANS AND MARINE ORGANISMS

The oceans contain plankton, the foundation or basis of the marine food chain. These microscopic organisms in the ocean are necessary for marine life. Increases of UV-B rays resulting from decreased ozone cause serious problems for plankton in metabolism and photosynthesis — necessary occurrences for life. Further, UV-B causes mutations in certain marine organisms and harms the eggs of others.

The ozone depletion occurring in Antarctica could undermine the entire food chain. All sea creatures depend in some way on plankton. Disturbing our ozone layer causes significant quantities of ultraviolet radiation to pass through the weakened ozone shield and destroy plant cells. The ozone hole over the Antarctic is allowing large doses of ultraviolet radiation to reach the plankton for 6-8 weeks every year. This increase in radiation is destroying it. Without the plankton, our food chain collapses. Further, with ozone depletion occurring, life everywhere is at risk, not just life in the Antarctic.

Plankton also absorbs a significant amount of atmospheric CO_2. It plays an important role in balancing CO_2 levels in our atmosphere.

EFFECT ON CLIMATE

A loss of ozone in the stratosphere will result in cooler temperatures in that layer of our atmosphere. Cooler temperatures in the stratosphere may mean altering global wind patterns, which in turn will change our weather. Since this is all new to us, no one is exactly sure what will happen to the climate.

AN OZONE HOLE IS DISCOVERED

In 1982, a British scientific team found a seasonal ozone hole over the continent of Antarctica. This surprised scientists and climatologists from all over the world. In fact, by October, 1987, and again in 1989, ozone concentrations over Antarctica had reached their lowest values ever. More than 50% of the ozone had disappeared over the continent. The ozone hole was the size of the United States and taller than Mount Everest.

Most atmospheric scientists were confident during the 1970's and 1980's that they had a good understanding of our atmosphere. When the United States launched the weather satellite Nimbus 7 in 1978, scientists programmed the ozone measuring instruments to ignore "low readings" of ozone as an error. Past readings of historical data proved to scientists what the accepted ozone levels should be. If the levels were below these historical measurements, the instruments on Nimbus 7 ignored the low readings. This was exactly what took place in the late 1970's and early 1980's. The satellite simply ignored low ozone readings over Antarctica. Scientists finally found the low readings by land-based observation instruments in 1982.

It was the first time since their work began in 1957, that they found very low ozone readings. At first the team thought their equipment may have been faulty. The scientists replaced key equipment and in 1983 and 1984 continued their ozone reading investigations. The group confirmed their low ozone readings in 1984. This may be perhaps the most important atmospheric discovery ever made.

A positive confirmation of the ozone loss in the hole over Antarctica occurred in 1987. More than 150 scientists and support personnel from 4 nations formed the Airborne Antarctic Ozone Experiment. Two specially equipped planes actually flew into the ozone hole. What they found astounded the entire scientific community. Ozone levels registered less than one-half of historic worldwide ozone levels. The ozone hole had 50% less ozone than normal and in some areas of the large hole, ozone had disappeared completely. Scientists scrambled to figure out what was happening and when the jury came in, the verdict was unanimous. Man-made compounds of chlorine from industrial products were causing the ozone destruction. Nearly all the chlorine comes from man-made CFC's.

The ozone hole usually appears annually during September and October. The hole lasts about 4-6 weeks. The temporary ozone hole, over 2,000 miles across, typically covers the entire continent of Antarctica. The ozone depletion occurs in the layer of atmosphere from about 12 to 20 miles up.

The failure to predict this ozone hole has raised many questions for the scientific community. No one had ever predicted such an occurrence, and people were beginning to ask what else could happen. Indeed, how much abuse can our atmosphere take before it starts breaking down?

OZONE DEPLETION IS MAN-CAUSED

Scientists now attribute nearly all ozone destruction to releases of chlorine from industrial products (CFC's). Chlorine rises to the stratosphere and goes through a chemical reaction there. Cold temperatures coupled with clouds and sunlight help chlorine destroy ozone. Evidence now shows an ozone hole forming over the Arctic. Temperatures are slightly warmer in the Arctic than in Antarctica. For this reason we may not encounter large ozone destruction over the Arctic. If ozone depletion continues in the Arctic, the results would be much more severe since most of the planet's population lives in the northern hemisphere. Ozone losses in the Arctic now range upwards of 5-10%. Ozone also tends to shift around in the stratosphere. This could cause the

Antarctic Ozone Loss for Month of October

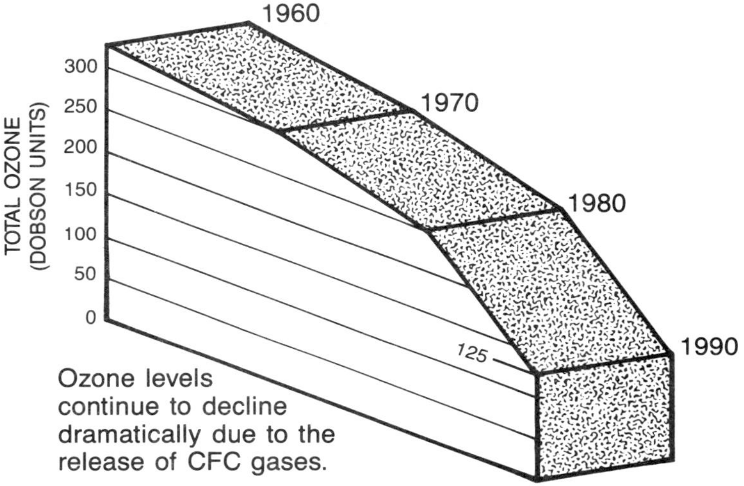

Ozone levels continue to decline dramatically due to the release of CFC gases.

entire ozone layer in the northern hemisphere to decrease as the ozone begins to even out and adjust to regaining a balance. During the winter of 1989, scientists noted a 5% reduction of ozone over much of the entire northern hemisphere. What has already happened in Antarctica may soon occur in the Arctic.

However, it is still not clear exactly how much ozone depletion is occurring, although we do know that our ozone layer is decreasing. One estimate from satellite data shows an ozone depletion of 3% from 1978-1984. In 1988, ozone decreases in the northern hemisphere were 1.7-3% and registered 5-6% in 1989. There is difficulty in measuring ozone concentrations since ozone constantly shifts around in our atmosphere, and this also makes comparisons difficult. Because the ozone hole over the South Pole area cannot get any deeper, there is now widespread fear that the hole will spread outward and affect surrounding areas. In fact, in 1987, 3 of 5 ozone monitoring stations located in Australia registered very severe drops in ozone levels. These levels were the lowest on record for that time of year.

As mentioned in the first chapter, the Greenhouse Factor is the critical threshold at which severe and unpredictable reactions will take place in our atmosphere. The Greenhouse Factor was at work, and we did not know about it. In the 1930's, atmospheric scientists knew of four basic chemical atmospheric reactions. Today, scientists know of hundreds of chemical reactions that occur in our atmosphere and the list continues to grow. The large hole caught us off guard and surprised scientists. Suddenly, in 1982, we had an ozone hole the size of the United States and the scientific community was at a loss to explain why.

Now that we are aware that CFC's destroy ozone, we should take steps to limit CFC production. Yet, an increase in CFC production continues throughout the world. Further, the most popular types of CFC's (Freon 11 and 12) are very stable in the atmosphere. They remain there for about 100 years.

The good news is that we have found the cause of the ozone problem. The bad news is that CFC's endure 100 years and there may be other chemical reactions taking place, or about to take place, of which we are unaware. Chemicals released years ago may just now be reacting with other chemicals in the atmosphere, some effects of which are unknown. Another problem is that CFC's are cumulative in our atmosphere. Prior releases of CFC's are just now reaching the stratosphere, since it takes about 6 years for the CFC's to reach that level.

We have above us, the largest, longest, and certainly the most dangerous chemistry experiment ever performed in our atmosphere. Only time will tell what havoc has been wrought in our skies.

THE GLOBAL CHEMISTRY EXPERIMENT ON OUR ATMOSPHERE CONTINUES

It is difficult to measure the total amount of chemical gases and other toxic pollutants released into our atmosphere. The EPA conducted the first United States nationwide survey of air pollution in 1987. In fact, it took a congressional mandate to perform the study. The EPA released results of the 1987 U.S. air pollution survey in 1989. The survey revealed that industry released

1,350,000 tons of 308 different toxic chemicals into the atmosphere. United States law regulates only 8 of these toxic chemicals. Sixty of these chemicals cause cancer. The study showed the principal causes of air pollution in the United States are automobiles, utility plants and factories.

The 1,350,000 tons, or 2.7 billion pounds, of toxic chemicals that enter the atmosphere each year include industrial and manufacturing releases. What the survey does not include is the 55.7 million tons of carbon monoxide released into our air. The primary source of the gas is the more than 140 million automobiles in the United States. Carbon monoxide, a deadly gas, is a by-product of combustion from car engines. Automobiles account for over three-fourths of all carbon monoxide emissions in the United States.

The 308 toxic chemicals plus carbon monoxide are not the only releases. There are also over 18.5 million tons of sulfur dioxide, 17.7 million tons of nitrogen dioxide, 19.6 million tons of ozone-producing gases and 7.3 million tons of lead-related gases released into the air annually.

The continued daily release of these large amounts of gases into our atmosphere has started to cause significant imbalances up above. Higher temperatures from a pronounced greenhouse effect coupled with global ozone losses and the seasonal ozone hole, are the symptoms. We are crossing the threshold of the Greenhouse Factor.

Chapter 4

WORLD CLIMATE IN CRISIS: The Greenhouse Influence

Major causes of recent climate changes are indeed directly attributable to one item: man's pollution of our skies. Disturbing the delicate balance of our atmosphere has a direct influence on climate. Historically, as relatively minor changes in the makeup of the atmosphere have occurred, drastic and sometimes cataclysmic changes in the climate on earth have occurred.

The greenhouse effect left ignored and unchecked not only will cause warmer temperatures, but will change climate patterns. This will dramatically alter the way in which we live. It would be the first time in history that whole civilizations alter their lifestyles from something other than political or social upheaval. The new nature caused by man will result in disasters brought on by changes in the weather and eventually the entire climate.

The greenhouse influence on our weather patterns is very real. We have precious little information on exactly what could happen to our weather, because this increased greenhouse effect and ozone loss is so new to us. We've never had this much global pollution and atmospheric disturbance. Scientists and climatologists are unable to model weather and climate accurately. Our atmosphere is much too complex for even the most sophisticated of computers. However, science is providing us with new evidence. Heat-trapping pollutants released by man, along with the destruction of the ozone layer, will create unprecedented changes in climate patterns.

The Greenhouse Factor has in fact happened before. For example, the dinosaurs disappeared from the face of the earth after being the dominant life form for 140 million years. Dinosaurs were mostly vegetarians and their extinction almost certainly was the result of an interruption or reduction of their food. Somehow the climate changed and curtailed the growth, or even killed, the plants on which they fed. All signs point to a relatively sudden and dramatic cooling of the planet. In this case, the Greenhouse Factor came about by the impact on our planet by a comet or asteroid or by massive volcanic eruptions. Either of these happenings would have ejected enough debris into the atmosphere to block enough sunlight to cool the planet. This cooling would have caused the polar ice caps to expand. Sea level would have lowered; and with the colder temperatures and more dry land exposed, rainfall would have decreased. One other theory holds that wildfire may have also engulfed the planet, killing even more plants and animals. Some or all of these factors combined to inhibit or wipe out the plants on which the dinosaurs fed. With the food supply declining, the extinction of the dinosaurs was imminent.

The Greenhouse Factor was, in this case, a severe and sudden disturbance of our atmosphere which caused the planet to cool. It was impossible to predict or control. The mass extinction of many species, such as the dinosaurs, was due to nature. Man may cause the next Greenhouse Factor.

Global air temperatures have risen about 1° F in the last 100 years. This doesn't sound significant, but put in historical perspective, it is alarming. The last one degree of warming took 2,000 years. Put another way, our planet has warmed by only about 2° F since the birth of Christ. Half of that change took place just since the invention of the telephone.

The consequences of a warmer world are already affecting weather and climate patterns. The disturbance of our atmosphere is causing an impact on our climate. Since 1982, ocean surfaces have warmed by about 1° F. Why is this significant? Warmer bodies of water generate more intense storms such as hurricanes and typhoons. To add to this, a redistribution of ozone in the stratosphere will alter the temperature of the stratosphere. This event will cause an impact on weather patterns around the globe.

To understand the consequences resulting from the Greenhouse Factor, let's look at some basics of weather and climate.

WEATHER

If you were to ask children or adults for a definition of the weather, just about all of them would blurt out words like "rain," or "clouds," or "storms." They're right, of course, but not completely. The weather is the condition of the sky. It can be any combination of the above or, with bright sunshine, none of the above. A hurricane, a gentle shower, a dust storm, or clear skies are just a few examples of the spectrum of atmospheric conditions called weather.

"Weather" is short term. "Weather" is what is happening outside in snapshot form. Weather is not as simple as the forecast you hear on the radio or see on television.

Weather starts with the sun. The radiation from the sun travels 93 million miles before striking our earth's atmosphere. Lucky for us, we have the ozone layer to filter out most of the ultraviolet radiation or our weather wouldn't matter much. The earth is fortunate enough to have an atmosphere that filters out the worst of the sun's rays. Our atmosphere also helps hold in the heat at night. Without it, we would have Hell on earth during the day and worse than Arctic cold at night.

Virtually all of our weather occurs in the lower atmosphere. This thin skin of air surrounding us we call the troposphere. The troposphere contains about 85% of our atmosphere's mass. The next layer of the atmosphere, the stratosphere, extends up to about 30 miles. It is cold and steady by comparison. Only an occasional thunderstorm penetrates briefly and leaves a thin veil of ice crystal clouds. The air moves about, but you can't see it without the clouds to track this air movement. The air is thin, cold, dry and fragile. Change occurs more slowly in the stratosphere than close to the earth. Without the benefit of moisture, the air this high up is slower to cleanse itself of pollution. The ozone layer we refer to lies within the stratosphere. The stratosphere contains nearly 15% of our atmosphere's mass. Above the stratosphere, from 30 to 50 miles, lies the mesosphere and above 50 miles, on the edge of outer space, the thermosphere. No weather occurs in the mesosphere or the thermosphere, so they have virtually no effect on our climate.

When sunlight reaches the ground, it warms the land and oceans. Water evaporates from the oceans and lakes in the form of invisible water vapor. As water vapor cools, it condenses into visible moisture we call clouds. When the air becomes saturated (100% humidity) any further condensation causes rain or snow. When atmospheric conditions are right, thunderstorms can develop. Thunderstorms account for much of the severe weather reported in the United States. Thunderstorms can cause 100 mile-per-hour winds, torrential rains and even hail. A thunderstorm is also the first ingredient necessary to spawn a tornado, the most severe local storm on our planet.

In the Tropics, the moisture-laden air rises and forms showers that can drop heavy rains. The oceans in the Tropics account for the largest share of the storage of the sun's energy. This makes the tropical areas very important in what happens to the earth's climate.

The Tropics are also where hurricanes mature. A weather disturbance in the stratosphere will

Diagram of the Layers of the Atmosphere

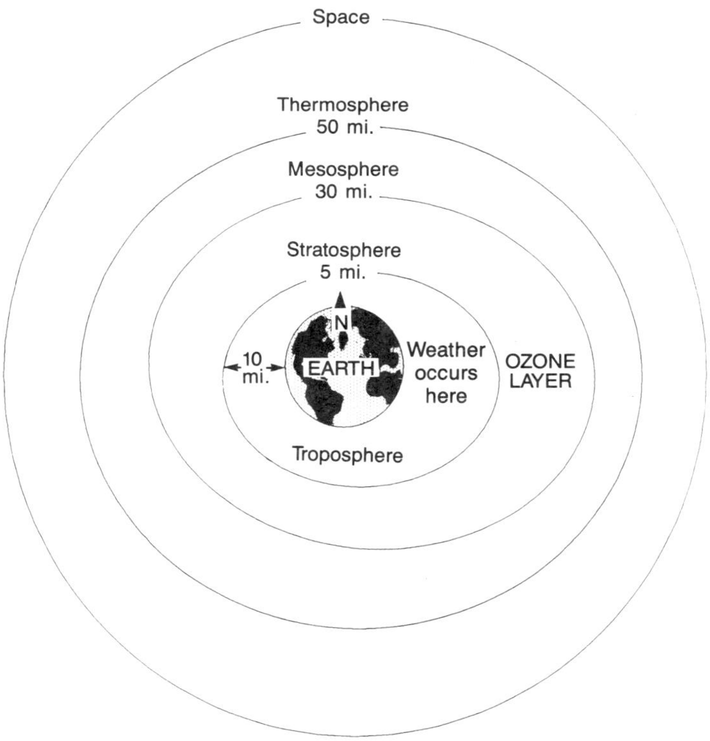

sometimes enhance a cluster of Tropical thunderstorms. If the water is warm enough, this storm can "feed" off the warm water. The storm continues to grow from a tropical depression to a tropical storm. If the growth persists, and the sustained winds in the storm reach 75 miles-per-hour, we call it a hurricane. Hurricanes are massive storms with their cloud bands extending for hundreds of miles in all directions. They are extremely destructive when they reach populated areas. Hurricanes cause wind damage, torrential rains and towering waves of water that can inundate large areas. Sometimes entire islands disappear.

This collection of factors is what makes up the changes that we hear, see and read about daily. These conditions and many more are what we call weather.

CLIMATE

Climate is the weather over a long period. Climate really is the average weather. Collecting weather data over a long period, usually decades, for a specific area, provides the inclination of the weather for that area. The word climate is from the Greek *klima* meaning inclination. When you ask someone what the weather is like in a given area, they are more likely to describe the climate. For example, if you live in a desert area you might say, "It's hot and dry here." Compared to other areas, that is true. Check the weather records.

You'll find that it can be cold and wet sometimes — but it is "inclined" to be hot.

WHAT MAKES CLIMATE

Location primarily determines climate. The climate is warmer near the equator and colder near the poles. It is also usually warmer at lower elevations than on mountaintops. When the location is near the ocean, the climate probably has more rain and a smaller range in temperature. The rain is easy to explain when you are near water since there is naturally more moisture in the air. However, temperature requires a little more explanation. A large body of water, such as an ocean, can store a large amount of heat energy. Oceans warm up slowly and also cool slowly. This tends to smooth out large differences in temperature. The same principle applies to a swimming pool but on a much smaller scale. It always takes the water extra time to warm up in the early part of the swimming season in spite of several hot days in a row. On the other hand, we usually have extra days of swimming late in the season even after some days of cooler temperatures. Thus, climates near the ocean have smaller temperature variations.

We already mentioned that the distance from the equator is another major factor in climate. Even in the winter, the sun is strong at the equator. That is why the climate is warm at the equator and cold at the poles where the sun's radiation is weakest.

Our earth is always turning on its axis and that revolution, once every 24 hours, is what causes the change from day to night. If we weren't rotating on our axis, the climate would be much different. There would always be warm air rising from the sunny side of the earth, and there would always be cold air descending on the dark side. This difference in temperatures would set up very distinct climates on either side. They wouldn't be pleasant. The sunny side would have hot temperatures and high humidity because of the oceans evaporating rapidly. This in turn would lead to clouds, rain, and probably widespread intense storms. The other side of the planet would be dark, cold and dry. Ice might even develop in the oceans. Climate would vary little at a given location. Fortunately, thanks to the rotation of the earth, this isn't what happens.

JET STREAM

When the earth rotates on its axis, this complicates the otherwise simple air currents. The spinning earth deflects the cold air flowing from the poles to the equator, making the jet stream. This is a high speed core of winds that flows from west to east in the higher latitudes. This prevailing flow of winds we call the polar jet stream or just jet stream. For simplicity, you can think of this band of winds as the "storm track." Most storms and weather systems will parallel the flow of the jet stream. Also, it can be thought of as the dividing zone between the colder polar air to the north and the warmer tropical air to the south. The jet stream does not stay in a fixed position but shifts from north to south. The jet stream is primarily responsible for day to day changes in the weather.

A much weaker jet stream, called the "subtropical jet stream," lies closer to the equator. It typically flows from east to west and is responsible for most of the weather occurrences in the tropical areas.

The earth is not exactly perpendicular to the sun's rays. As we revolve around the sun, the amount of energy varies during the year. This yearly variation is what we call the seasons.

So climate is actually very complex. Local effects also add to the complexity of our climate. Vegetation, nearby geographic features and other factors can all have important effects on our local climate.

WHAT CHANGES CLIMATE

We've talked about the fairly obvious effects on climate, and we have known about those for a long time. As science has advanced, we have learned more about factors that change climate and are less obvious.

Factors that change climate fall into three groups: extraterrestrial, natural climate changes and human-caused climate changes.

EXTRATERRESTRIAL CLIMATE CHANGES

The sun is the main cause of extraterrestrial change in our climate. Although we think of the sun as constant, sensitive measurements of the output of the sun show that it varies. One of the variations you may have heard about is the sunspot cycle. This not only involves a cycle in the number of sunspots, but also a cycle in the intensity of the energy from the sun. This low point in the sun's energy causes a temporary cooling of the earth's climate.

The sunspots about disappeared between the years 1645 and 1715. The decrease in the energy from the sun resulted in devastating winters in Europe and in America during this period. This period of climate we call the "Little Ice Age." In fact, several theories put forth in the late 1980's speculate that increased solar activity has recently masked ozone depletion. This masking happens since more solar activity will produce more ozone than normal without an increase in solar activity. Around 1991-1992, when solar activity decreases, ozone levels may fall dramatically.

Astronomers have observed sunspots for almost as long as we have had telescopes. The connection between sunspot cycles and climate is much more recent. The number of sunspots ranges from a maximum to a minimum in about 11 years; so a complete cycle is around 22 years. The sunspot cycle is significant to our climate. Many climate forecasters, such as those for *The Old Farmer's Almanac,* rely heavily on solar activity to make their yearly predictions.

Comets may be another cause of extraterrestrial climate change. As mentioned earlier, a comet or some other extraterrestrial body struck earth. The impact ejected large volumes of comet, or earth debris, into the atmosphere. As this dust and debris fanned out in the atmosphere, it acted like a sunscreen and blocked a good portion of the energy from the sun. This in turn cooled the earth's climate. The resulting drastic change in the vegetation patterns on earth caused the extinction of the dinosaurs, who were unable to adapt quickly enough to the changes.

If an asteroid or comet should strike one of earth's oceans, the effects would be different. Billions of gallons of sea water would vaporize. This would cause a pronounced greenhouse effect, shrouding the planet in clouds. Some of the sun's rays would get through and warm the earth, but the heat wouldn't escape. This would cause a change in plant life and cause mass extinctions.

NATURAL CLIMATE CHANGES

The idea that dust in the atmosphere can cause lower temperatures at ground level is quite old. Benjamin Franklin speculated that the cold winter of 1783-1784 was caused by the dust in the upper atmosphere. His theory was right.

In 1816, volcanic dust caused a worldwide short-term climate change. Newspapers ran stories of the cold spring and summer next to follow-up stories of a monumental natural disaster. No one made the obvious connection. Dust in the upper atmosphere caused the climate to cool so much that we now call 1816 the "Year Without a Summer."

THE YEAR WITHOUT A SUMMER

In 1815, an immense volcanic eruption and explosion in what is now Indonesia literally erased Mount Tambora on the island of Sumbawa. This volcano reduced Mount Tambora in height by 4,200 feet and displaced an estimated 25 cubic miles of debris. Much of the debris ended up in the atmosphere as fine volcanic dust. On the island of Java, three hundred miles away, the sky became overcast at noon from the ash cloud. Showers of ash covered houses, streets and fields to a depth of several inches. Even at this distance from the disaster, residents heard the explosions at intervals, sounding like distant thunder or artillery.

In the months following the eruption, the dust ejected by Mount Tambora began to circle the globe. As the dust fanned out, it became a thin layer in the stratosphere and began to cause dramatic changes in the weather. In May, observers in New England began to comment on the lateness of the spring frosts. In June, a series of blustery cold waves deposited 3 to 6 inches of snow that damaged or destroyed the remaining crops in northern New England. July brought another killing frost. Two more frosts in August destroyed all but the hardiest of crops and vegetables. Crop prices soared to high levels. Some cattle starved because farmers couldn't afford the exorbitant prices brought about by the hay and grain shortages. The price of hay in New Hampshire, in a few instances, skyrocketed from the more normal $30 per ton to as high as $180 per ton.

Some farmers gave up after the devastating crop failures of 1816. The number of people moving west doubled in the years 1816 and 1817, compared to other years in that decade.

With the Napoleonic Wars and poor crops, Europe was even harder hit than the United States. Food shortages were particularly severe in some European cities. Soldiers in the Loire Valley in France had to protect grain carts from looters so the supplies would make their way to market.

Famine in India weakened enough people to give rise to a local cholera epidemic in Bengal. Historians believe that the British military operations spread the disease to the Afghans and Nepalese. It spread from there to the Caspian Sea, then slowly westward to the Volga and to the Middle East. Without planes and trains, the disease spread slowly. There is good evidence that by 1832 the same outbreak of the disease had spread to New York. By July of 1832, 100 people a day were dying of cholera in New York City. No doubt the epidemic would have spread around the world anyway, but the famine in India almost certainly accelerated the process.

A volcanic eruption, half a world away, caused failed crops on several continents, near riots in Europe and aggravated a global epidemic. It was the "Year Without a Summer," and the year of the phrase "1800 and Froze to Death."

El NIÑO

The El Niño/Southern Oscillation is a natural air/sea interaction measured in the Pacific Ocean between Australia and the west coast of South America. El Niño refers to warm surface waters and the opposite, La Niña, refers to cool surface waters off that same coast of South America. The warmer waters of El Niño disrupt the weather by shifting or even reversing air circulation patterns. This cyclic interaction has an immediate impact on the fishing industry in South America. It also has far-reaching effects on the climate of both the Northern and Southern Hemispheres.

In 1982 and 1983, waters off the Western coast of South America warmed by about 3-5° F. El Niño spawned storms causing severe flooding in South America. In the United States, the California coast experienced some of the worst storms ever recorded.

MAN-CAUSED CLIMATE CHANGES

Though not as dramatic as a volcano or comet, humans on this planet are slowly changing the climate. We release noxious gases into the air, denude the rain forests, and these in turn are creating upper atmospheric disturbances. Changing atmospheric temperatures brought about by greenhouse gases and atmospheric pollution will change our climate.

WARMER TEMPERATURES CAUSE MORE TROPICAL STORMS

Continued release of greenhouse gases into our skies is causing the natural greenhouse effect to warm. The natural greenhouse effect provides us with a warm blanket of air. Without it, we could not have life on earth. Many computer projections show our atmosphere warming and predict continued warming by about 1° F per decade. The warming of our atmosphere also heats our oceans. Oceans cover nearly three-fourths of the earth, and are thus the dominant feature of our planet, significantly affecting the climate.

A 1° F average increase in temperature may generate hurricanes that are one-third stronger. If the oceans warm further, the hurricane season, now 6 months long, will last longer. Hurricanes will also become more frequent and more intense. This will be a direct result of man's warming the atmosphere by releasing more and more greenhouse gases which trap heat.

As oceans warm, the hurricane-threatened area will spread northward up the eastern coast of the United States and Canada. Warmer air and ocean temperatures cause increased evaporation over the oceans. The increased evaporation provides more moisture and energy for spawning and sustaining hurricanes.

Warmer seas will also disrupt global air and ocean circulation patterns. As the air and ocean circulations change, weather patterns will change. Since these changes interact with one another, they are very complex and difficult to predict.

DROUGHTS

The warming of our atmosphere will also cause more evaporation to take place, bringing severe droughts. Droughts occur when high pressure persists over an area. High pressure in the atmosphere usually brings clear skies and warmer than normal temperatures. Droughts are more serious when they occur during the growing season in our agricultural regions.

1988 was the warmest year on record. During the summer of 1988, the United States experienced the worst drought recorded since the 1930's. For the first time in many years, it was necessary to use crop surpluses from earlier years to meet the food demand in our country.

DEFORESTATION

Compounding climate change is the problem of deforestation. Our tropical rain forest, creator of so much of the oxygen we breath, is in trouble. Slash and burn agricultural techniques are being used to clear the rain forest. The Amazon Forest in Brazil is an area about two-thirds the size of the United States. More than 12% of that forest is already gone. Tropical rain forests hold a critical link in our atmosphere's balance. This is actually a two-fold problem. The first part of the problem is the more obvious. A destroyed forest can no longer absorb CO_2 and, through the process of photosynthesis, give off oxygen. The second part of the problem concerns the methods used to cut down trees. Burning is the favored method of clearing these forests in most areas. This releases all the CO_2 that the plant had stored in its fiber which in turn, adds to the greenhouse problem.

Another concern stemming from deforestation in the rain forests is the extinction of animal and insect life. The rain forest is home to thousands of species of animals and insects. When we destroy the rain forest we destroy their habitat. It is thought that there are many undiscovered species of animals and insects residing in the rain forest. The destruction of the rain forest may prevent their discovery.

HEAT ISLANDS

Viewing our planet from up in space on a clear night is very interesting. Over the United States, one sees scattered bright spots, farther apart in the west, more concentrated along the east coast. It looks like islands of bright light on a dark sea.

The bright spots, the cities, are islands of heat. This so-called heat island effect is another way we change our climate. The change in temperatures in urban areas is significant. It is common for a city to be 15° to 20° F warmer than the surrounding countryside.

Rapidly growing metropolitan areas have experienced astonishing temperature rises in a short span of years. For example, in the Phoenix, Arizona metro area, the average daily temperatures jumped as much as 6° F in the last 30 years. This is comparing the 30-year normals ending in 1970 with the 30-year normals ending in 1980. The change was slight for daytime temperatures, but some of the changes at night were astounding. Throughout the years, normal morning low temperatures rose 3° to 6° F. Further increases in temperature will occur as urban areas continue to grow in the 1990's.

Studies show that wind patterns around cities change when temperature patterns change. The wind difference may not be noticeable, but the wind pattern changes may influence new rain and snowfall patterns. More studies will shed light on this possibility.

Precipitation patterns change downwind of some urban areas. Exhaustive studies of the rain and snowfall patterns around the St. Louis metro area confirm that rainfall increases in the opposite direction of the prevailing wind. This may be due in part to the rising heat from the city causing more clouds and in turn more rainfall. Heavy industry in the metro area contributes to the particles in the air. These particles make it easier for clouds and precipitation to develop. There is no doubt that cities are influencing, as well as changing, our local climate.

OZONE LOSS AFFECTS CLIMATE

A loss of ozone in the stratosphere will mean cooler temperatures in that part of the atmosphere. The cooling up above could mean significant changes in global wind patterns. If global wind patterns change, significant changes in local weather may occur.

Our fast-food, air-conditioned lifestyle has led us to be the major producers of chlorofluorocarbons (CFC's). These complex chemicals are destroyers of the ozone layer high in the atmosphere. As these gases reach the ozone layer, they are broken into simpler elements that destroy ozone. The discovery of the "hole" in the ozone over Antarctica has alerted us to the global nature of this problem.

Air-conditioners, especially in automobiles, are large consumers of CFC's. Leaks in the systems and careless discharge of chemical coolants are major causes of CFC releases into the atmosphere. Foam cups and containers, such as those we get at fast-food restaurants, are also ozone destroyers.

Climate will also change as we continue to pollute our skies with the greenhouse gases. These gases trap the reflected heat off our planet and create the greenhouse effect. It is difficult to stop the active forces which are causing climate changes. The energy foundation of our global economy burns oil, gas, and other fossil fuels. This burning emits carbon dioxide (CO_2) which, in turn, traps heat in our atmosphere. Hotter temperatures will cause more heat waves, more drought conditions, less rainfall and rising sea levels.

Nature has changed our climate with comets, asteroids and volcanoes. We are now changing our climate, not by cataclysm, but by increments. When we add up all the increments, do the pluses and minuses total zero? Or, do the increments total widespread and unpredictable change of the climate — The Greenhouse Factor?

Chapter 5

WHY SHOULD WE CARE
What Can We Do?

Continued releases of chemical pollutants into the atmosphere are causing two important events to take place. First, certain gases, such as carbon dioxide (CO_2) and methane, increase the trapping of heat reflected from the earth. This causes the atmosphere to warm. Secondly, certain pollutants, such as chlorofluorocarbons (CFC's), are destroying the ozone layer. This allows more ultraviolet radiation to pass into the atmosphere and downward to the earth's surface.

Changes from increasing atmospheric pollution directly affect us. The changes have a significant impact on sea level, climate, vegetation, acid rain and human health.

SEA-LEVEL RISE

Sea-level rise is a serious threat. As the air surrounding our planet warms, the ocean water warms and expands. We call this thermal expansion. As the ocean waters warm, this will in turn begin to melt glaciers and the polar ice caps. The North and South Poles contain large reservoirs of water in the form of ice. Should the ice melt, lower-lying land areas around the world will flood. Complete melting of the ice caps and glaciers would raise the world's sea level by more than 200 feet.

The average sea-level increase over the past 7,000 years has been only about 1-2 inches per century. Thermal expansion has already raised sea levels about 4 inches since 1900. This expansion will cause about one-half of the future global sea level rise. Melting of the polar ice-caps will account for the rest of the sea-level rise.

One-third of the world's population lives within 40 miles of a coastline. Sea-level rise would affect the entire 12,383 mile coastline of the United States.

For example, Miami, Florida is only a few feet above sea level. A 10 to 12 inch sea-level rise can wipe out 100-200 feet of beach front. A sea-level rise of 12-15 feet would inundate a large part of southern Florida.

Global Sea Level

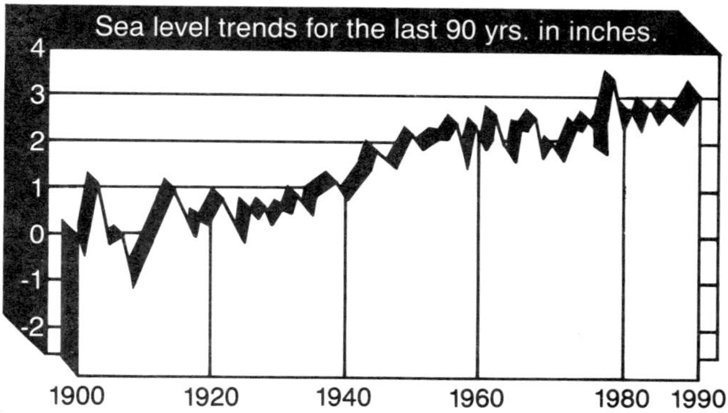

Recent sea-level rise is due in part to warmer temperatures causing a melting of glaciers and the polar ice caps.

Much of southern Louisiana, especially New Orleans, is at risk from just a slight rise in sea level. Most of metropolitan New Orleans lies below sea level. New Orleans already has a real threat from the advance of saltwater from the Gulf of Mexico. Pumps run continuously and levees contain the Mississippi to keep the city dry. A sea-level rise of just a few inches would bring New Orleans to the brink of disaster.

Since 1900, nearly a million acres of Louisiana land, including small off-shore islands, have disappeared. In only 8 years, southern Louisiana has sunk 4 inches due to the subsiding of soil (sediments) deposited from the Mississippi River. Sea-level rise will compound this already serious problem.

A rise in sea level would also affect other countries. In Holland, residents have reclaimed over 50% of the land area using sophisticated dikes and sand dunes to hold back sea water. In fact, more than 10 million Dutch people live on land that would be underwater without miraculous human-engineering feats.

Bangladesh and Egypt are two other countries affected by a slight sea-level rise. In these two countries, much of the population lives near the water. Bangladesh is in a flood-prone area with nearly all the land part of a low-lying river delta. In Egypt, the Nile Delta area is also at risk from a slight sea-level rise. Most populous cities are near or along the Nile River, a rich agricultural area and the center for Egypt's economic activities.

Sea-level rise can disrupt the entire economy in some countries. To add to this, entire population centers would need to move to higher ground. The next decade may very well bring a new breed of refugees. These people might be called environmental refugees.

WEATHER AND CLIMATE

Since 1982, ocean surfaces have warmed an average of nearly 1° F. Warmer air temperatures will raise the temperature of the ocean even more. This will cause more frequent and more severe storms. Higher seas, caused by a rise in sea level, will have a greater destructive impact on coastal areas.

Global warming will produce storms in the Tropics with up to 50% more energy and winds reaching more than 200 miles per hour. Tropical storms that reach hurricane-force winds now average about 6 per year. This number would surely increase with warmer ocean temperatures.

Warmer temperatures will also cause more droughts. Fertile farmlands will suffer drought conditions more frequently since these future droughts will be caused by man rather than periodic natural droughts.

In 1988 a devastating forest fire engulfed a large portion of Yellowstone National Park. More droughts will in turn result in drier conditions leading to more forest fires. This will increase the greenhouse effect even more because the burning of trees releases significant amounts of the heat-trapping gas, carbon dioxide.

WIND PATTERNS WILL CHANGE

With less ozone higher in the stratosphere where it belongs, more sunlight will pass to the lower stratosphere to manufacture ozone. This will cause a redistribution of ozone in the stratosphere. This redistribution of ozone will alter the temperature of the stratosphere and directly influence world climate. Colder air up above will change wind patterns. Since this is all so new to us, we don't know exactly what changes will occur in our climate.

HOTTER TEMPERATURES AND LIFE

Man has been able to adapt to living in a warmer world. Inventions such as air conditioning and evaporative coolers have made life tolerable in hot areas. However, not everyone in the world is lucky enough to live in air-conditioned comfort. Prolonged warmer temperatures will have a pronounced effect on lifestyles and even productivity. High temperatures during major heat waves cause many deaths. A warmer world means even more heat waves and more fatalities.

EFFECT ON LIVESTOCK

Man will not be the only one affected by warmer temperatures. Livestock, such as dairy cows, will also suffer. In order for livestock to keep cool, they will use more of their energy to fight the heat. Dairy cows, for example, will produce less milk. The cow will use more of its nutrients to keep cool rather than producing milk. This is only one example, but it points out that global warming has wide ramifications to man as well as livestock.

OUR CROPS ARE AT RISK

The amount of pollution in the air, combined with a change in climate, is directly affecting our vegetation. Air pollution is causing nationwide damage or death to many types of plants. Ground-level ozone damages chlorophyll necessary for photosynthesis.

An immediate concern is the damage being done to our crops. The severe drought in 1988 resulted in a 30 to 40 percent reduction in the United States' corn, soybean and wheat production.

Cotton, beans, peas and melons are especially susceptible to the harmful UV radiation. With less ozone above, more harmful UV rays will reach our crops. The radiation disrupts the plants' leaf cells, lowering their resistance to disease and insect attacks. Increased UV radiation can also affect the genetic structure of crops.

ACID RAIN

Another serious problem in our atmosphere is acid rain. The two primary pollutants that cause acid rain are sulfur dioxide and nitrogen oxides. Sulfur dioxide comes primarily from electric utilities burning high-sulfur coal. Electric utility plants and motor vehicles produce nitrogen oxides. Motor vehicles are responsible for almost 50% of the nitrogen oxide releases in the United States. Once in the air, sulfur dioxide and nitrogen oxides combine with other atmospheric chemicals to form sulfuric and nitric acids. When these chemicals mix with rain or snow, they fall to earth as acid rain.

Some of the rain can be as acidic as lemon juice. When the acid rain falls to earth it mixes with soil, poisoning some plants. Sulfur and nitrogen compounds in the rainfall combine to foster changes in the soil's chemistry. The affected soil diminishes a tree's efficiency in absorbing necessary nutrients.

Lakes and rivers also suffer from acid rain. Lakes can get so acidic they can't naturally cope with all the acid. The lakes begin to lose many types of organisms on which fish feed. Fish soon die off since they have less to eat and most cannot survive with the acidity in the lakes. This was the case in the Adirondack Mountains in upstate New York. One study estimated that nearly one-third of the lakes in this pristine preserve could no longer support life. The reason is the high acidity from the accumulation of years of acid rainfall. Acid rain is a local, regional and national problem. Acid rain clouds can travel more than 2,000 miles. Research shows that some acid rain in the northeast comes from midwest power plants' emissions from the burning of high-sulfur coal.

The problem of acid rain affects human health as well as the environment. Acid rain largely affects the respiratory system. Symptoms include chest tightness, emphysema and chronic bronchitis.

OZONE DEPLETION AND HUMAN HEALTH

Destroying the ozone layer is hazardous to our health. A depleted ozone layer means increasing ultraviolet (UV) rays reaching the earth. The effects of an increase in UV rays are many and include: more skin cancers, increase in cataracts and damage to the human immune system.

GROUND-LEVEL OZONE

Ground-level ozone is the major ingredient of smog. Most ground-level ozone comes from complex reactions of sunlight and automobile exhaust. Deadly ozone continues to increase at ground level, impairing breathing, and harming health in general.

Northern states experience high ozone levels from May through October while the southern states can have high ozone levels all year. A record number of American cities exceed the EPA safe air standard for ozone. In 1988, 75 cities showed ozone levels that were at least 25% above the EPA .120 parts per million (PPM) "safe level." New York City ozone levels can reach .200 PPM on heavily-polluted days. In California, ozone levels exceed the safe level for about half the year and often reach levels over .300 PPM. This is nearly 3 times greater than the EPA "safe level."

WHAT CAN WE DO?

ON THE INDIVIDUAL LEVEL

There are many ways in which we as individuals can combat ozone destruction and global warming. It won't be easy. If we were to practice environmental safety starting tomorrow, it would take at least ten years before the problems might begin to correct themselves. Even worse, the chlorofluorocarbons (CFC's) we release into the atmosphere today will remain there for about 100 years before deteriorating.

In the wake of global ozone destruction and ever-increasing temperatures, people must concern themselves with learning about the environment, as well as how to preserve it.

The battle against these problems begins with the individual. The size and scope of the problem leads many people to believe there is nothing they can do to make a difference. This is not the case. There are many simple ways in which each of us can help to make a difference.

TO PROTECT THE OZONE LAYER AT HOME OR AWAY

■ Avoid foam products containing or made with chlorofluorocarbons (CFC's) such as polystyrene (styrofoam). CFC's destroy ozone. Cups and plates made with polystyrene may also be around 500 years from now. Americans throw away over 25 billion styrofoam cups each year. Polystyrene also floats in our rivers and lakes. It resembles food as it crumbles and clogs the digestive system of marine life and other animals.

- □ **One** CFC-chlorine molecule can destroy as many as **100,000** ozone molecules over its 100-year lifetime.
- □ Avoid foam packaging that many fast-food restaurants still use.
- □ Buy paper picnic plates, cups and bowls — not styrofoam products made with CFC's.
- □ Buy eggs, meat and other products in paperboard trays.
- □ Patronize fast-food restaurants that wrap their burgers with paper products.

■ When shopping, don't buy household products containing chlorofluorocarbons (CFC's) or halons.

□ Always check the ingredients for CFC's in spray cans (aerosols). These include cleaning products, pesticides and other common household items.

□ Check the label for any of the following names (or similar names):

CFC-11 or Trichlorofluoromethane
CFC-12 or Dichlorodifluoromethane
CFC-113 or Trichlorofluoroethane
CFC-114 or Dichlorotetrafluoroethane
CFC-115 or Monochloropentafluoroethane

■ Use a plastic or metal cooler for camping or picnics.

□ It will last longer.

□ It won't release CFC's when it crumbles or you throw it away.

■ Avoid home insulation that contains CFC's.

□ Avoid CFC-blown or sprayed insulation such as urethane and isocyanurate foams.

□ Use non-CFC foam insulation such as fiberboard, fiberglass or cellulose insulation.

■ Don't discard that old refrigerator.

□ Refrigerators account for about 7% of total CFC and halon emissions.

□ Refrigerators contain CFC's in their coils and even their walls. Once the CFC escapes from the refrigerator, it enters the atmosphere and destroys ozone.

□ Call your local utility or recycling center to find out if they have a safe disposal program in place for refrigerators.

AT WORK

■ Avoid using polystyrene (styrofoam) cups at work.

□ Use a paper cup at the water cooler.

□ Bring your own coffee cup or mug to work.

■ Order and use "recycled" paper when possible.

■ Avoid using "packing peanuts" that contain CFC's.

□ Tell your mail room supervisor to replace the "peanuts" with paper-based packing materials.

□ Many form-fit packing materials also contain CFC's.

■ About 10% of aerosol cans still use CFC's as a propellant. Check the labels.

□ Many cleaning sprays and electronic parts cleaners contain CFC's.

□ Paints and solvents contain CFC's too. Avoid products with CFC-113 or methyl chloroform. Both destroy ozone.

■ Do not buy fire extinguishers that use halons. Halons have 10 times the ozone-destruction potential of CFC's.

YOUR VEHICLE

■ Car air conditioners contain CFC's. We commonly call the CFC's by the trade name of Freon. They represent over 25% of total CFC use in the United States. More than 65% of United States' cars have air conditioning. United States' cars emit more than 90 million pounds of CFC gases every year.

□ Fix leaky air conditioners immediately.

□ Replace your car air-conditioner hose every 3 years. Run your car air conditioner several times during the winter to keep hoses from drying out and cracking.

□ Patronize service stations that use refrigerant reclaiming systems. The service station pumps the refrigerant (CFC's) out of the compressor, purifies it, and recycles it.

□ If you really want to do something about ozone destruction, buy a car that does not have air conditioning.

TO HELP AGAINST THE GREENHOUSE EFFECT

■ Plant a tree. Trees absorb carbon dioxide.

□ A mature tree can absorb about 10-15 pounds of carbon dioxide each year. This helps reduce the over-abundant amount of carbon dioxide in our atmosphere that is causing warmer temperatures.

□ A tree can also shade homes and buildings and help reduce the need for air conditioning. Remember that air conditioners are a large user of electricity. Most electricity generation comes from the burning of fossil fuels at utility plants, which create carbon dioxide emissions. Less electricity used means that less carbon dioxide will enter our atmosphere.

□ Trees also serve to remove more pollutants from the air.

■ Practice car efficiency. There are over 400 million vehicles in the world. The United States has the most vehicles of any nation—over 140 million. Yet the United States has only 5% of the world's population. On average, every car emits about 5 tons of carbon dioxide (CO_2) annually. Of all oil used in the United States, nearly two-thirds is for transportation uses.

□ Buy a fuel-efficient car. That is a car that gets 30-50 miles per gallon (MPG). One gallon of gas burned produces about 20 pounds of CO_2. Less fuel burned means less emissions of the heat-trapping gas CO_2. A car that gets 18 MPG produces over 50 tons of CO_2 during its lifetime while a car getting 25 MPG produces less than 35 tons.

□ Reduce gas consumption: avoid fast starts;

Sources of Carbon Dioxide(CO_2) from Fossil Fuels

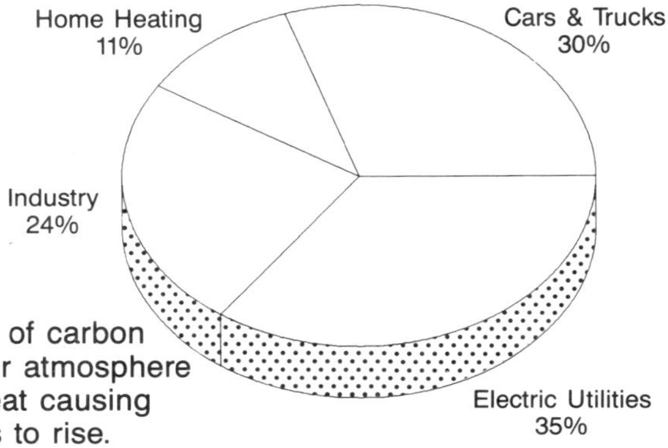

Rising levels of carbon dioxide in our atmosphere trap more heat causing temperatures to rise.

keep your engine tuned-up; keep your tires properly inflated, and limit the use of your air conditioner.

□ City driving consumes twice as much fuel as highway driving so avoid short or unnecessary trips.

□ Use public transportation, car pool or ride a bike.

■ Conserve energy at home and save money too. Over 11% of United States' carbon dioxide (CO_2) emissions are from energy used in the home.

□ Turn off lights when you don't need them. It will save you money and your utility company won't have to burn additional fossil fuels like oil or gas to generate electricity. This reduces the amount of carbon dioxide released into the air. Remember also that fossil fuels are a large contributor to acid rain.

□ Natural gas produces less CO_2 than either coal or oil. Use natural gas if you have the option.

□ If you own an oil furnace, have it cleaned once a year. It will run more efficiently.

□ Make sure your home insulation (non-CFC insulation) is adequate for the climate where you live. Invest in weatherstripping. Have a home energy audit performed by your local utility company to find ways to save energy and money.

□ Don't run the hot water unless you need to.

□ Install an insulator around your water heater.

□ Replace air filters regularly and you'll use less energy to heat and cool your home.

□ Set your thermostat a few degrees lower in the winter and try a higher setting with your air conditioner in the summer.

■ Install low-flow shower heads. These can reduce hot water flow by as much as 50%. This will save energy and money. Using less hot water requires less burning of fossil fuels to run your water heater.

■ Don't let water run when you're not using it such as when you brush your teeth.

■ Lighting consumes more than 20% of energy used in the home.

- □ Use fluorescent lights. These are 3 times more energy efficient than the incandescent type. One 40-watt fluorescent bulb gives off more light than 3 60-watt incandescent bulbs.
- □ Fluorescent bulbs will last much longer (8-15 times as long).
- □ You'll save energy, money and help curtail CO_2 emissions.

■ Buy a heat pump for efficient heating.

■ Buy an energy efficient refrigerator with a power-saver switch.

■ Buy a solar water heater for your hot water needs.

RECYCLING HELPS

■ Separate your trash and recycle your glass, cans, plastic and newspapers. Each ton of paper recycled saves almost 17 trees. Recycling a single run of a very large city Sunday newspaper could save over 50,000 trees.

- □ Recycling these items is a safe way of disposing of this refuse.
- □ There are recycling centers in most communities.
- □ Recycling means less energy needed to produce the same product.

■ Recycle your aluminum cans. Producing aluminum is both expensive and polluting. It is also one of the easiest metals for industry to melt down and recycle.

■ Buy products packaged in recycled or recyclable materials such as cardboard or glass—not plastic. Buy products that come in refillable containers.

■ At the grocery store, use paper bags, not plastic. If you are more ambitious, you can shop with your own reusable bags.

TO HELP CURTAIL POLLUTION

■ Have your car tuned up regularly. Ground-level ozone is at record levels.

■ Curtail unnecessary driving and use public transportation.

■ Use less energy at home. Turn off your lights. The less energy you use, the less need there is to burn fossil fuels. Less electricity generated by the utility results in less atmospheric pollution from an oil or gas-fired utility plant.

■ Use electric-powered garden tools rather than gas-powered weed trimmers and lawn mowers.

■ If you use a wood stove, use an energy-efficient model. There are about 10 million wood-burning stoves in use in the United States today.

■ Don't waste energy in your daily lifestyle.

■ Practice recycling.

What We Throw Away

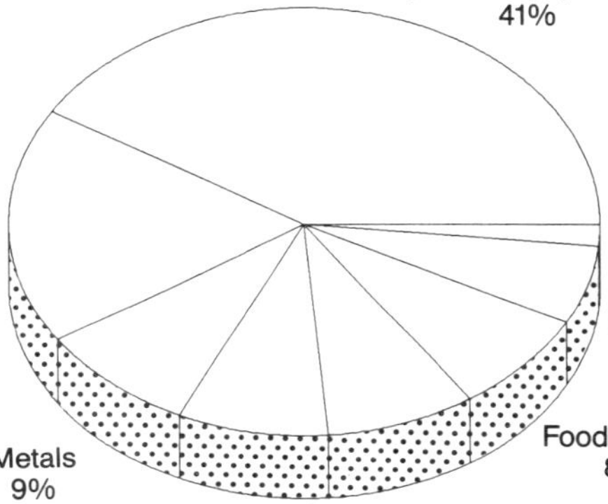

Recycling of household items such as newspapers and bottles saves large amounts of energy.

POLITICAL ACTION:
Write Your Representative

An individual can help reduce the global warming and ozone problems through political action. Political action is necessary to reverse the problems that now exist due to our throw-away lifestyles.

Individuals can act. They can write letters or form committees to lobby for better environmental concerns.

■ Write your Representative, Senator or local government official.

- □ It takes little time and only a postage stamp.
- □ Nearly all elected officials answer their

mail (re-election is important to them).

□ Letters are influential and government officials know letters represent the views of many who don't write. A courteous and unemotional letter is the most effective.

□ Let them know you support issues such as: a total ban on CFC's; increasing the MPG for cars; support of solar, wind and other non-fossil fuel sources of energy; or support for mass transit improvement.

□ Try to focus on one issue per letter, and if possible, identify a particular piece of legislation.

Only with the help of government action will our society move toward combatting environmental problems.

ON THE GOVERNMENTAL LEVEL

Many individuals are becoming environmentally responsible. Governments now need to be responsible in helping to curtail the problems of global warming, ozone destruction and air pollution. We need federal leadership.

Many argue that more scientific certainty on the greenhouse and ozone problems is needed before Congress can implement more environmental legislation. Others argue that while Congress waits for more scientific facts and proof, the more climate change and ozone damage will occur.

Let's take a look at the efforts of the United States and local governments to help curb the damage to our atmosphere. Although the atmospheric problems have been with us for a long time, politicians are just beginning to take serious steps to control this problem.

THE CLEAN AIR ACT OF 1970

In 1970, the United States Congress passed the Clean Air Act. The Clean Air Act directed the Environmental Protection Agency (EPA) to set up health or air standards for 6 common air pollutants. These are sulfur dioxide, particulates, carbon monoxide, nitrogen dioxides, ozone and lead. The EPA set standards to reduce pollution from industry and automobiles. Auto emission standards from the Clean Air Act forced car manufacturers to build cleaner cars. Due to these regulations, new cars were 6 times cleaner than those produced in the late 1960's. The implementation of the catalytic converter was instrumental in the success of cleaner-burning cars. The catalytic converter converts many (but not all) pollutants to less harmful exhaust gases. Auto-related pollutants actually decreased by the mid 1980's versus the early 1970's. The government mandated less polluting auto engines. Eliminating lead from most fuels was the first step. Without the lead, cars could have catalytic converters which dramatically cleaned up the exhaust from the auto engines.

However, by the late 1980's, air quality began once again to deteriorate. Cities reported many violations of EPA standards between 1986 and 1988 due to automobile pollution. The number of unsafe or violation days from carbon monoxide or ozone pollution quadrupled between 1986 and 1988 in the United States. By 1988, over 110 million Americans lived in communities where smog or auto pollution exceeded the EPA's safe level.

There have been amendments to this Act, such as some adopted in 1977, that actually strengthened the Act. However, the present Act is only a first step. It was not until 1987 that the EPA conducted the first pollution survey of toxic gases released by industry. The results showed annual releases by industry of over 1.3 million tons of 308 various toxic chemicals. Further, the original 1970 Clean Air Act did not address the problem of acid rain. Congress did have good intentions in 1970 by passing the Clean Air Act. However, the 20-year-old Act needs updating as widespread abuse on our atmosphere continues.

The task of strengthening the 1970 Clean Air Act is not an easy one. State governments squabble over who should pay and how much to spend.

The major provisions addressed by the new clean air legislation are:

- □ Curbing acid rain by reducing sulfur

dioxide emissions from coal-burning power plants.

□ Tighter automobile tailpipe emission standards to combat urban smog.

□ Industry curtailment of certain toxic chemicals. Since 1970, only 8 pollutants are now under control. Currently, there are 308 known toxic pollutants entering our air.

AIR POLLUTION REDUCTION

Our government does a good job of monitoring air pollution such as ground-level ozone. Sadly, monitoring is only one step and does nothing to stop production of air pollution. It just tells us how bad, or good, the air quality is.

Some ways to reduce air pollution such as gound-level ozone (the major part of smog) include:

□ Curtailing automobile emissions

□ Reducing smokestack emissions from industry

□ Installing vapor recovery systems on gas pumps at service stations

□ Implementing mass-transit programs.

□ Encouraging the use of alternative fuels such as methanol and ethanol.

GOVERNMENT AND OZONE

The 1978 United States Ban on CFC's in Aerosols

In 1978, the United States Government banned CFC's from most, but not all, aerosol cans. A

1974 theory that showed a link between ozone destruction and CFC's in the stratosphere prompted this action. Banning of CFC's as aerosol propellants also took place in Canada and Sweden in the 1970's. Despite government intervention, CFC use remains widespread and today spray cans are a large source of CFC emissions.

The United States' ban was an excellent first step toward curbing CFC use but stops short of eliminating the problem. CFC use is widespread in many products such as air conditioners, certain household chemical sprays and foam containers. Despite United States restrictions on certain CFC-aerosol uses, the United States accounts for almost one-third of global CFC consumption. Americans use almost 6 times more CFC's than the global average. World production of CFC's is 2.5 billion pounds annually.

The problem with ending CFC use altogether is economic. Most CFC substitutes are less effective and much more expensive to manufacture. Until cheaper substitutes make it to the marketplace, economic issues win over the environmental repercussions.

ELEVATING THE ENVIRONMENTAL PROTECTION AGENCY (EPA) TO A CABINET LEVEL POSITION

Elevating the EPA to a cabinet level position may be a good move by the federal government.

This might allow the EPA to achieve the recognition it needs to carry out environmentally responsible programs. A cabinet level position would place environment on an equal basis with other important government departments. Theoretically, the environment would then receive the same attention as defense, transportation, commerce and the other cabinet departments.

On the other hand, increased EPA spending and new regulations may not be the answer to the problems of a polluted planet. Many claim that environmentalism is an excuse for the government to move back into managing our lives. The United Nations argues for a free market approach to environmental issues.

STATE AND LOCAL GOVERNMENTS CAN SET A GOOD EXAMPLE

State and local governments can act to address environmental problems much faster than the federal government. State and local governments are now at work passing and invoking significant regional legislation. This is much more difficult to accomplish on the Federal level. Successful local environmental legislation is working in the areas of air pollution, ozone depletion and recycling.

- In 1989, Vermont passed one of the nation's toughest state-level measures to stop ozone destruction. The state bans CFC use in all car air conditioners starting in 1993.

□ Beginning in 1993, state governments in New York, New Jersey, Massachusetts, Maine, Vermont, Connecticut, Rhode Island and New Hampshire will mandate a 50% decrease of allowable auto pollution.

□ Many states have a bottle law. This law requires a refundable deposit on bottles. This monetary incentive provides good motivation for people to recycle.

□ California has perhaps the most aggressive anti-pollution campaign in the world. The plan calls for a radical reduction in air pollution that will cause lifestyle changes. The California plan has 123 measures banning or restricting polluting practices and chemicals. The 20-year, 5,500-page plan calls for such measures as the banning of all aerosol hair sprays and deodorants (not just those containing CFC's). It also forces employers to encourage car pooling and requires companies to install anti-smog equipment regardless of cost.

POLITICAL PARTY ACTION

With only 5% of the world's population, the United States consumes about 25% of the world's energy. The United States is a chief contributor to global warming, yet it is doing little to fight the battle on a governmental level. In Western Europe, however, environmental concerns are a high priority. In the late 1970's, the Green Party political movement began. These environmental (or Green Party) political entities have a strong foothold at the governmental level. The Green Party has members in over 15 countries and even members in parliaments of 8 countries. These political groups have been successful in directing government monies, programs and legislation toward lessening environmental problems and concerns.

THE MONTREAL PROTOCOL ON CFC's

In 1987, 24 nations signed a document known as the "Montreal Protocol." The purpose of this legislation was to reduce consumption of CFC's. Since then, over 60 nations have agreed to sign or have signed this agreement. The "Montreal Protocol" is the first global effort to curb production of the ozone destroying chemicals known as CFC's. The treaty calls for a 50% reduction in the production and use of CFC's by the year 2000.

Delaying governmental action on banning all CFC use prolongs the ozone crisis. It takes a CFC molecule about 6 years to rise through the atmosphere before it reaches the "ozone layer" in the stratosphere. Also, the CFC's have a lifetime in our atmosphere of about 100 years.

Environmental groups have also moved forward to do something about the ozone problem. This has prompted industry to act. Dupont, the world's largest producer of CFC's, announced a significant program to develop safe alternatives. Nissan says it will stop using CFC's in automobile and other air-conditioners by 1993. McDonald's has agreed to stop using food containers manufactured with CFC's.

This progress is good news but the bad news is that CFC use continues to grow by about 4% annually. Also, the Montreal Protocol does not cover all types of ozone-destroying CFC's. There is concern over the enforcement of the treaty. Lastly, there are many countries who have not signed the Montreal Protocol and have no intention of doing so. CFC production persists and our ozone layer continues to disappear.

TOWARD GLOBAL COOPERATION

Greenhouse warming and ozone destruction are not just local or national problems. They are cause for global environmental concern. To combat the problems, more international agreements are necessary across all geographic boundaries. Many developing nations concern themselves with immediate problems like feeding, clothing and housing their people. They are not thinking about environmental issues. Most developing countries would find it difficult to make cutbacks in their industries that depend on fossil fuels. Cutting back on fossil fuels to help lessen the greenhouse effect would cause considerable economic hardship. Estimates are that 95% of all future population growth will be in developing

U.S. Energy Sources

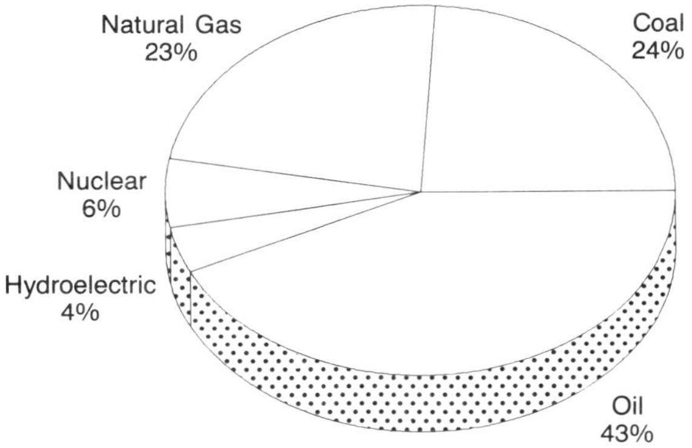

The United States is heavily dependent upon fossil fuels such as oil, coal and natural gas for energy uses. Fossil fuels release carbon dioxide which in turn traps more heat in the atmosphere.

countries. There is a need for new energy supplies to support the growing population in these developing countries.

Many developing countries are heavily dependent upon the burning of fossil fuels for energy. It's not easy to change entire economies overnight. Hydroelectric energy is a safe and clean alternative to fossil fuels but hydropower cannot meet all electricity demands. One alternative is the use of nuclear energy. Nuclear does not pollute the atmosphere but produces toxic waste. Countries just view the use of fossil fuels as the lesser of the evils.

Individual countries cannot go it alone and need help from others more capable. This is not an easy task. Many countries do not have political relations with each other. In fact, it's difficult for the International Olympic Committee to get all countries to participate in sports once every 4 years. There is always dissent, protest and boycotts from certain countries just over sporting events. Agreement on the world environment is going to be much more difficult.

There have been international steps to address global warming and ozone destruction. Performance of much of the work has been outside the United States, especially in Europe.

- In March, 1989, English Prime Minister Thatcher hosted a 3-day international conference on ozone destruction. Her aim was to convince more countries to sign the Montreal Protocol. By the close of the conference, 20 more countries agreed to sign.
- In March, 1989, the governments of Norway, France, and the Netherlands hosted a meeting to consider global warming and ozone destruction. Representatives from 24 countries attended. The result was The Hague Declaration. Among other items, it recommended that the United Nations have enforcement powers over environmental compliance.
- Great Britain has proposed that the United Nations Security Council have world environmental authority.

□ The Soviets have recommended an ecological council at the United Nations.
□ In July, 1989, France's President Mitterand focused his annual summit meeting almost exclusively on environmental issues.
□ One Australian environmental plan calls for the planting of 1 billion trees during the 1990's to help curtail global warming.

A NEED FOR ACTION BY GOVERNMENTS

Countries around the world must agree to take steps to curtail global warming and ozone depletion. Action must include reducing the amounts of greenhouse gases and ozone-destroying chemicals that enter our atmosphere. The issues at hand are many, but here are the key items that need attention:

□ Cut use of fossil fuels.
□ Halt destruction of forests, especially rain forests.
□ Move toward renewable energy sources such as solar, wind, photovoltaic, and geothermal. (Over 95 countries now use wind to create power.)
□ Curtail annual production of the over 2.5 billion pounds of ozone-destroying CFC's.
□ Find effective replacements for CFC's.
□ Reformulate the gasoline used in vehicles to cut harmful emissions.

THE GLOBAL CHEMISTRY EXPERIMENT IN OUR ATMOSPHERE CONTINUES

Each of us can help ease the crisis

We have learned that the atmosphere is disposable. There is a crisis in our atmosphere. Increasing greenhouse gases are warming our planet and the protective ozone layer is disappearing. We also have learned that we can help to reverse the process. Just by driving our cars less and using less electricity we can reduce the burning of fossil fuels. This will help curtail the increase in CO_2 released into our atmosphere. Eventually, this will ease the global warming problem.

Some gases released into our atmosphere are not leaving. As in the case of CFC's, those gases will be in our atmosphere for the next 100 years destroying ozone and trapping heat.

An economical replacement for CFC's must be found soon. This will reduce the destruction of the ozone layer in the future. However, we have to act right now. CFC's released today take 6 years to get to the stratosphere. The ozone destruction cannot end as long as we continue to release CFC's into the air.

The crisis in the atmosphere is probably reversible if we begin to take action now. If we wait, we will cross the critical threshold of severe and unpredictable reactions in the earth's atmosphere. We can prevent the Greenhouse Factor from happening. We have abused our atmosphere for over a century now and it has, so far, been very forgiving. The atmosphere probably has the capacity to recover from our past abuses if we work together to end the abuse of our skies. We must take corrective steps to remedy the situation now.

About The Author

Ed Phillips has always had an interest in the atmosphere and weather. He received his degree in Aeronautical Meteorology in 1973 from Parks College of Aeronautical Technology of St. Louis University. During college, Ed worked in radio as a weather broadcaster. After college, Ed moved to Arizona and began television weathercasting in 1979, while continuing to work in radio as an on-air meteorologist. In 1981, Ed was granted a Seal of Approval from the American Meteorological Society.

Ed Phillips' weather almanacs, published annually since 1980, are well known throughout Arizona. The 1990 KTAR Radio Edition has over 50,000 copies in print.

As an author, broadcaster and frequent guest speaker on the environment, atmosphere and weather, Ed has received numerous awards for his work. These include the Literacy in Media Award in 1987; First Place Award for special broadcast coverage from Arizona's Associated Press Broadcasters Association; and in 1985 Ed was chosen as a member of the AMS Board of Broadcast Meteorology.

An active pilot, Ed is a member of the Board of Directors of the Arizona Pilots Association and Scottsdale Airport Commission. In 1990, Ed was chosen as President of the national Air Awareness Council. Ed currently resides in Scottsdale, Arizona, with his wife Anne and son Alex.

APPENDIX I

REGIONAL INFLUENCES OF THE GREENHOUSE EFFECT

The Warming Atmosphere

A doubling of CO_2 in the atmosphere during the next 50 years will produce a rise in the mean temperatures of the earth by 3-9° F. Computer models of the atmosphere forecast this rise to be 2-3 times more at the Poles and higher latitudes than in the Tropics. Those models also predict that the Arctic (North Pole) will be significantly more affected than the Antarctic (South Pole).

Imagine the effects of that much warming at the Poles. The ice caps would begin to melt thereby releasing volumes of water into the oceans. This would, in turn, cause a rise in the sea level around the world. Estimates in the sea-level rise are from 1 to 6.5 feet in the first 100 years and up to 15 feet in 300 years. A sea-level rise of 15 feet would be catastrophic. That rise in sea level would affect eleven million people living in coastal areas in the United States alone. In the world, 30% of the total population of 5 billion live within 40 miles of an ocean. That means that sea-level rise would affect one and one-half billion people. This is a staggering toll only tempered by the knowledge that it will probably occur over a few hundred years.

The atmospheric warming by the increased greenhouse effect will have profound effects even without considering the oceans. Desert areas in the mid-latitudes will probably get hotter. The heat will likely spread deeper into the continents of Asia and North America intensifying the effects of low rainfall.

A warmer atmosphere will mean the Polar Jet Stream will stay closer to the Poles year-round. Therefore, the rain and snow producing storms steered by the Jet Stream will stay at higher latitudes. This will cause the more regular rainfall patterns to shift farther north.

It is also possible we will have sharper contrasts in the temperature of the air masses on both sides of the jet stream. This will make for strong cold fronts which in the warmer months may spawn more severe thunderstorms and tornadoes. Since the storm track will be farther north, the severe weather will also shift farther north. Increased snowfall at the higher latitudes may be a wintertime effect. This will allow some very cold air masses to develop over the snow cover. The jet stream will help hold these masses of air farther north longer, intensifying the cold. So in spite of the global warming, some continental areas (again North America and Asia) may have record outbreaks of winter cold.

Likewise, in the warmer months of the year, hot, dry air masses will persist over the continents longer and cause record hot spells.

The El Niño/Southern Oscillation is a widely studied and reported air/sea interaction. This phenomenon causes unusually warm water to occur periodically off the coast of South America. This warm surface water has a dramatic effect on the amount of plankton in the water on which fish feed. As a result, bird populations decrease and the fishing industry has smaller catches. El Niño affects only a portion of the Pacific Ocean. Warmer world oceans would cause shifts in seafood production and changes in the food chain that would be incalculable.

A warmer atmosphere will also hold more water vapor. In the atmospheric temperature range of 32° F to 86° F, the capacity to hold water vapor doubles for each 18° F rise in temperature. The parts of the globe with the largest increases in temperature, the Poles and higher latitudes, will have significantly more moisture in the atmosphere. This will lead to increased clouds, rain and snowfall. Mid-latitude regions far from the oceans will likely record a general decrease in clouds and rainfall. Examples would be on the Asian Continent and the middle of North America. More evaporation in the Tropics will cause more clouds and rainfall too.

The increase in water vapor and clouds will also reflect more of the sun's energy back into space. This will actually help decrease the greenhouse warming.

The Warming Oceans

Water expands as it warms. The global atmospheric warming will cause the oceans to warm and expand slightly. This will account for an even greater rise in the sea level. However, this effect will be minor in comparison to the much greater effect of the melting polar ice caps on the rise in sea level.

The warming of the oceans has another effect. Hurricanes only form over the warm waters of our planet. Imagine what would happen if the waters of the oceans were much warmer and the area of warm water much greater. Hurricane meteorologists estimate that the increase in ocean temperatures will cause an increase in hurricane strength by one-third. The number of storms will also increase because the warm water spawns them more easily. A one-third increase in strength will mean some of the worst hurricanes will have sustained winds of over 200 miles per hour. Storms of that intensity will cause catastrophic damage and storm surges that exceed 20 feet. These waves would inundate coastal areas that have never been close to flooding from previous storms.

Global warming and ocean warming will also cause more evaporation. Water evaporating from the oceans will help slow the rise in sea level, but only slightly. Of more importance, warmer water in lakes and inland seas will evaporate more readily. This will actually **lower** the levels of those bodies of water where rainfall remains the same or less.

GLOBAL WARMING AND YOUR AREA

West Coast

Sea level rises will flood coastal areas and cause monumental problems in major coastal cities. The San Francisco Bay Area will be hard hit because so many acres of land are just above the current sea level. Protecting sea level property (such as San Francisco International Airport) will be very expensive.

Warmer temperatures in the atmosphere, and warmer oceans, will probably mean a dryer and warmer climate for southern California. This will add to the high fire danger in so many of the heavily populated canyons around Los Angeles. With the higher temperatures, air pollution in the Los Angeles Basin will worsen. Ground-level ozone will increase even further. The lack of rain and fewer Pacific weather systems will mean longer air pollution episodes.

Fewer Pacific storms will also affect seasonal rainfall farther north along the west coast. Summers will be dryer and warmer, while Spring and Fall will be wetter. The winters will probably be wetter, too, and that will mean more snowpack on the higher coastal mountains. Cities around Puget Sound will have to contend with the problems of protecting property from the rising sea level.

Southwest Deserts

The Southwest already experiences the highest temperatures in the country. The global warming will add to those already high numbers and probably set some astounding temperature records. Phoenix, Arizona is already the hottest major metropolitan area in the country. Summer temperatures exceed 100° F for 85 days each year on average. The all-time record is 122° F (June 26, 1990). As the global warming progresses, the record high in Phoenix could climb by 3° to 7°.

Spring, early summer and fall will probably get hotter and dryer. Winters will be even milder, making the season even more appealing to winter visitors. Summer is the "wet season" in the desert and, surprisingly, it may get wetter. The strong daytime heat of summer causes air to rise over the desert. By mid to late summer, the air that replaces this air has origins in the Tropics. Higher temperatures would cause even greater heating and the need for more replacement air. Since the Tropical waters will be warmer, this air will be farther north than in the past. Therefore, the Tropical air will more easily flow into the southwest. This will prolong the wet season with more summer thunderstorms and greater rainfall. This potential summer increase in rain, however, may not offset the decrease in rainfall during the rest of the year.

Mountain States

The shift of the jet stream northward will play a major role in the climate of the Mountain States. Winter snowpack in the Rockies is crucial to the Mountain States. Unfortunately, it will probably be lower as the greenhouse effect increases. The jet stream will shift farther north taking winter storms with it. The more reliable yearly snowpack will move more into the Canadian Rockies and out of the United States Rockies. Summers will probably be warmer in the Mountain States. That will increase water usage. This will tighten the already short water supply from the reduced winter snowpack.

Great Plains/Midwest

Lack of rainfall in the middle of the country can have disastrous results on our food supply. We had a sample taste of this during the Depression years. Lack of rainfall and improper farming methods plunged part of the Great Plains into the "Dust Bowl." Winds blew the dry topsoil great distances uninterrupted by trees or hedgerows.

The rains finally returned, ending one of the worst long-term droughts in our history. Agricultural production returned to some of the Dust Bowl after the drought due to conservation-minded farming methods. Other "Dust Bowl" areas returned to production by the use of ground water irrigation.

The greenhouse effect will likely result in less rainfall for the middle of the nation. Obviously, this will have drastic affects on the rain-dependent farming in this area. Lower rainfall will reduce crop yields and in some cases lead to total crop failures. Unusually hot temperatures, also likely to occur from the increased greenhouse effect, may accompany these droughts. This will worsen an already bad situation. Less winter snowfall is also possible. This will harm winter wheat and other crops dependent on this moisture.

On the positive side, the growing season will be longer. Marginal growing areas with an adequate water supply will have an increase in crop yields. Estimates are that for every 1.8° F rise in temperature, the corn belt will shift about 100 miles northeast. With the global warming expected, the corn belt will shift about 200 to 800 miles to the northeast. For this crop alone there will be major effects on the United States, while Canada will benefit from warmer climate and a much larger corn cash crop. Similar shifts are possible with other crops. The extent will depend on the drought tolerance of the crop and future adaptability of crops to climate change.

Gulf Coast

One of the wettest parts of the country already, the Gulf Coast may get wetter as the greenhouse effect increases. With warmer waters in the Gulf of Mexico and higher evaporation, more moisture will be available to develop clouds and rain. This part of the country will probably be the least affected by temperature changes, although with the warmer water, temperatures will also warm. The rise in sea level will be devastating to cities such as New Orleans. Most of that metropolitan area is already below sea level. Much of the Mississippi Delta will be under water causing massive losses to crops, homes and cities.

A good portion of the rainfall along the Gulf Coast comes from storms of tropical origin. Since these storms will be more frequent and stronger, rainfall in this area will probably be higher. This is one part of our country where a "super" hurricane may strike. The increasing greenhouse effect may spawn "super" hurricanes with winds over 200 miles per hour and a storm surge of 20 feet or more.

More rainfall may actually benefit some crops along the Gulf Coast, assuming they are outside areas susceptible to flooding.

Great Lakes

The Great Lakes region may stand to benefit as much from the greenhouse effect as any other part of the country. Shipping lanes that freeze over in winter may now stay open all year with the rise of temperatures. This part of the country may have less summer rainfall and more winter rains and snows. That could cause a net rise in the Great Lakes and hasten the cleansing of pollutants from the Great Lakes. However, higher lake levels will also cause more shoreline erosion and maybe some flooding along the shores of such cities as Detroit, Cleveland and Chicago. Cities subject to lake-effect heavy snows in the winter may have greater snowfall. The warmer lakes will have less ice to prevent the cold winter air masses from evaporating water from them. The moisture-laden air drops the heavy snow in the "snow belts" downwind of the lakes. Warmer water and warmer winter air masses will hold even more moisture. The "snow belts" will get even snowier. This is certainly not good news to residents of Buffalo, Cleveland and Erie, Pennsylvania.

East Coast

Except for rising sea levels, the effects of global warming will be less significant in much of the East. Rainfall patterns inland will be similar to the past. If anything, rain and snowfall will be greater due to the higher ocean temperatures and higher moisture content in the air. Temperatures will rise some, but not nearly as much as in the Great Plains and Midwest.

If the sea-level rise is as much as some predictions, the economic costs for the East Coast for the next two centuries will be immense. Major cities, such as New York and Boston, will probably try to build sea walls at a staggering cost. A civil engineering project of this size would take de-

Rising Sea Level

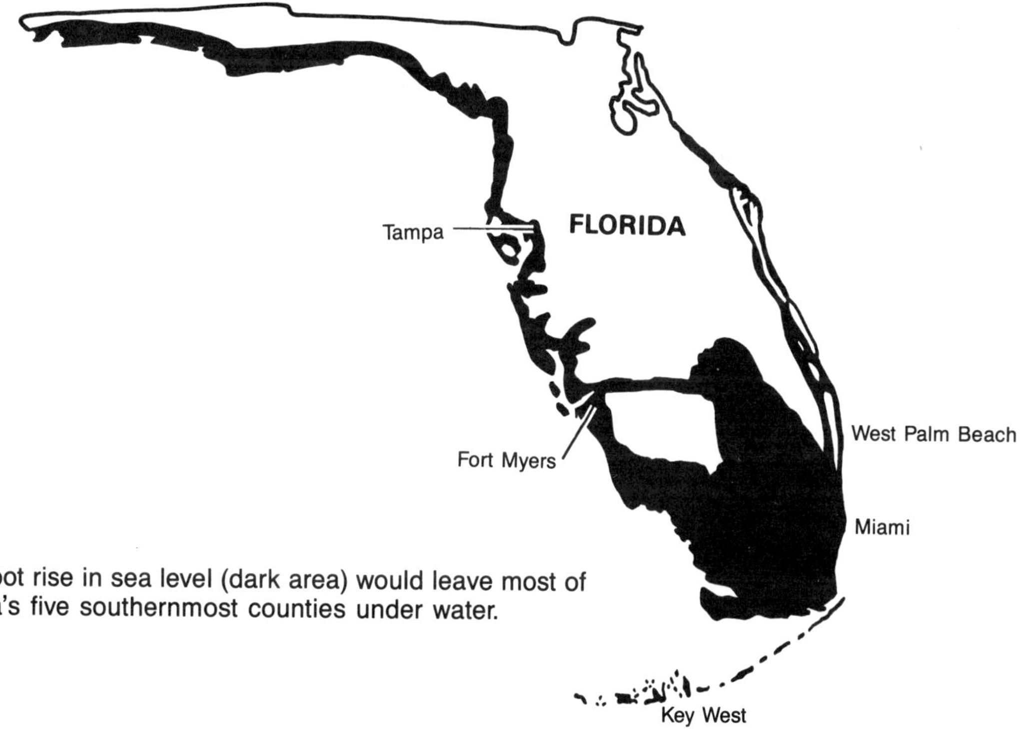

A 12-foot rise in sea level (dark area) would leave most of Florida's five southernmost counties under water.

cades to complete. The worst storms might breach the sea walls in spite of the most heroic efforts. The raising of docks and wharves, necessary for continued commerce, would be another staggering expense. Some of the finest beaches on the shores of New York and New Jersey will gradually disappear. The Outer Banks off the coast of North Carolina would be under water. The famous beaches like Miami Beach, Palm Beach and Daytona Beach would suffer the same fate. Even Cape Canaveral would be vulnerable to high tides and storms. The Florida Keys would disappear with a sea-level rise of 15 feet.

Residents of smaller cities and towns along the East Coast may abandon low-lying areas and move to higher ground. The economic results of fighting the rising sea level will probably force some towns to just disappear. Rebuilding may prove to be too costly.

Along the southeast coast and the east coast of Florida, some of the most devastating changes in our weather may occur. Due to warmer Atlantic Ocean temperatures, the hurricane season will be longer and the storms more frequent and intense. Stronger hurricanes will threaten the coasts with storm surges of 20 feet or more. This 20 foot surge is on top of the already higher sea level. Since the water will be warmer farther north, the chance of a major hurricane in cities such as Washington, D.C., New York City, and Boston will be greater.

Canada

While the United States will face monumental social and economic problems, the benefits of the global warming will probably outweigh the detriments in Canada. As mentioned earlier, the growing season will lengthen. The United States corn belt will move 200 to 800 miles northeast, depending on how much global warming occurs. Other agricultural crops will grow well in the Prairie Provinces of Canada, too. Indeed, Canada may become the breadbasket of North America, if not the world.

Warmer temperatures in Canada will probably cause an explosion of population growth. As more of the country becomes hospitable to agriculture, jobs will become available farther and farther north. Displaced American farmers may make up much of the population growth. The farmers may move north after years of low crop yields and little or no profit from their work.

The rising sea levels will put a lot of Canadian real estate under water. Much of the flooding will be in the sparsely populated areas of the country. However, Vancouver, Montreal and Quebec will have to cope with the same problems facing other cities near sea level.

APPENDIX II

TEMPERATURE RECORDS

Record High Temperatures
Major United States Cities
(Through 6/30/90)

Note: Cities in **Bold** have achieved an all-time high temperature in the 1980's — the warmest decade on record.

City	° F	Month/Year of Record
Akron, OH	**101**	**7/88**
Albany, NY	100	7/53
Albuquerque, NM	**105**	**6/80**
Amarillo, TX	108	6/53
Anchorage, AK	85	6/69
Asheville, NC	**100**	**8/83**
Atlanta, GA	**105**	**7/80**
Atlantic City, NJ	106	6/69
Austin, TX	109	7/54
Baltimore, MD	**105**	**8/83**
Baton Rouge, LA	103	6/54

Birmingham, AL	**106**	**7/80**
Bismarck, ND	109	7/73
Boise, ID	111	7/60
Boston, MA	102	7/77
Bridgeport, CT	103	7/57
Buffalo, NY	99	8/48
Burlington, VT	101	8/44
Charleston, SC	**104**	**7/86**
Charleston, WV	**104**	**7/88**
Charlotte, NC	104	9/54
Chattanooga, TN	106	7/52
Cheyenne, WY	100	6/54
Chicago, IL	**104**	**6/88**
Cincinnati, OH	**103**	**7/88**
Cleveland, OH	**104**	**6/88**
Colorado Springs, CO	100	6/54
Columbus, GA	**104**	**7/86**
Columbus, OH	102	6/44
Concord, NH	102	7/66
Corpus Christi, TX	104	7/39
Dallas-Fort Worth, TX	**113**	**6/80**
Dayton, OH	**102**	**6/88**
Denver, CO	104	7/39
Des Moines, IA	**108**	**8/83**
Detroit, MI	**104**	**6/88**
El Paso, TX	112	7/79

Eugene, OR	**108**	**8/81**
Fargo, ND	**106**	**7/88**
Fort Myers, FL	**103**	**6/81**
Fort Wayne, IN	**106**	**6/88**
Fresno, CA	**111**	**7/84**
Grand Rapids, MI	**100**	**7/88**
Greensboro, NC	**103**	**8/88**
Hartford, CT	102	7/66
Helena, MT	105	8/69
Honolulu, HI	**94**	**9/88**
Houston, TX	**107**	**8/80**
Huntsville, AL	**102**	**7/86**
Indianapolis, IN	104	7/54
Jackson, MS	**106**	**7/80**
Jacksonville, FL	105	7/42
Juneau, AK	90	7/75
Kansas City, MO	**110**	**8/84**
Knoxville, TN	103	9/54
Lansing, MI	**100**	**7/88**
Las Vegas, NV	**116**	**7/85**
Lexington, KY	**103**	**7/88**
Little Rock, AR	**112**	**7/86**
Lincoln, NE	**107**	**6/88**
Long Beach, CA	111	10/61
Los Angeles, CA	110	9/63
Louisville, KY	105	7/54

Lubbock, TX	**108**	**7/83**
Madison, WI	104	7/76
Memphis, TN	**108**	**7/80**
Miami, FL	**98**	**8/87**
Midland-Odessa, TX	**112**	**7/89**
Milwaukee, WI	**103**	**8/88**
Minneapolis, MN	**105**	**7/88**
Missoula, MT	105	7/73
Mobile, AL	104	7/52
Montgomery, AL	105	6/54
Nashville, TN	107	7/52
Newark, NJ	105	7/66
New Orleans, LA	**102**	**8/80**
New York, NY	106	7/36
Norfolk, VA	**104**	**8/80**
Oklahoma City, OK	**110**	**8/80**
Omaha, NE	114	7/36
Orlando, FL	102	5/45
Peoria, IL	**105**	**6/88**
Philadelphia, PA	104	7/66
Phoenix, AZ	122	6/90
Pittsburgh, PA	**103**	**7/88**
Portland, ME	103	8/75
Portland, OR	**107**	**8/81**
Providence, RI	104	8/75
Raleigh, NC	**105**	**8/88**

Rapid City, SD	**110**	**7/89**
Richmond, VA	105	7/77
Rochester, NY	100	6/53
Sacramento, CA	115	6/61
Saint Louis, MO	**107**	**8/84**
Salt Lake City, UT	107	7/60
San Antonio, TX	**108**	**8/86**
San Diego, CA	111	9/63
San Francisco, CA	106	6/61
Seattle, WA	**99**	**8/81**
Shreveport, LA	107	8/62
Sioux City, IA	**108**	**6/88**
Spokane, WA	108	8/61
Stockton, CA	114	7/72
Syracuse, NY	98	6/53
Tampa, FL	**99**	**6/85**
Toledo, OH	**104**	**6/88**
Topeka, KS	**110**	**8/84**
Tucson, AZ	117	6/90
Tulsa, OK	112	7/54
Washington, DC	**104**	**7/88**
Wichita, KS	113	7/54
Wilmington, DE	102	7/66

SOURCE: National Climatic Data Center (NOAA)
Local Climatological Data
Asheville, NC 28801

RECORD HIGH TEMPERATURES BY STATE

STATE	° F	LOCATION	DATE
Alabama	112	Centerville	September 5, 1925
Alaska	100	Fort Yukon	June 27, 1915
Arizona	127	Parker	July 7, 1905
Arkansas	120	Ozark	August 10, 1936
California	134	Death Valley	July 10, 1913
Colorado	118	Bennett	July 11, 1888
Connecticut	105	Waterbury	July 22, 1926
Delaware	110	Millsboro	July 21, 1930
D. C.	106	Washington	July 20, 1930
Florida	109	Monticello	June 29, 1931
Georgia	112	Louisville	July 24, 1952
Hawaii	100	Pahala	April 27, 1931
Idaho	118	Orofino	July 28, 1934
Illinois	117	East St. Louis	July 14, 1954
Indiana	116	Collegeville	July 14, 1936
Iowa	118	Keokuk	July 20, 1934
Kansas	121	near Alton	July 24, 1936
Kentucky	114	Greensburg	July 28, 1930
Louisiana	114	Plain Dealing	August 10, 1936
Maine	105	North Bridgton	July 10, 1911
Maryland	109	Cumberland & Frederick	July 10, 1936
Massachusetts	107	Chester & New Bedford	August 2, 1975
Michigan	112	Mio	July 13, 1936
Minnesota	114	Moorhead	July 6, 1936
Mississippi	115	Holly Springs	July 29, 1930

Missouri	118	Warsaw/Union	July 14, 1954
Montana	117	Medicine Lake	July 5, 1937
Nebraska	118	Minden	July 24, 1936
Nevada	122	Overton	June 23, 1954
New Hampshire	106	Nashua	July 4, 1911
New Jersey	110	Runyon	July 10, 1936
New Mexico	116	Orogrande	July 14, 1934
New York	108	Troy	July 22, 1926
North Carolina	109	Weldon	September 7, 1954
North Dakota	121	Steele	July 6, 1936
Ohio	113	near Gallipolis	July 21, 1934
Oklahoma	120	Tishomingo	July 26, 1943
Oregon	119	Pendleton	August 10, 1898
Pennsylvania	111	Phoenixville	July 10, 1936
Rhode Island	104	Providence	August 2, 1975
South Carolina	111	Camden	June 28, 1954
South Dakota	120	Gann Valley	July 5, 1936
Tennessee	113	Perryville	August 9, 1930
Texas	120	Seymour	August 12, 1936
Utah	116	Saint George	June 28, 1892
Vermont	105	Vernon	July 4, 1911
Virginia	110	Balcony Falls	July 15, 1954
Washington	118	Ice Harbor Dam	August 5, 1961
West Virginia	112	Martinsburg	July 10, 1936
Wisconsin	114	Wisconsin Dells	July 13, 1936
Wyoming	114	Basin	July 12, 1900

The last date of occurrence is printed. Some records also occurred at other locations and/or on earlier dates.

RECORD LOW TEMPERATURES BY STATE

STATE	° F	LOCATION	DATE
Alabama	- 27	New Market	January 30, 1966
Alaska	- 80	Prospect Creek Camp	January 23, 1971
Arizona	- 40	Hawley Lake	January 7, 1971
Arkansas	- 29	Pond	February 13, 1905
California	- 45	Boca	January 20, 1937
Colorado	- 60	Maybell	January 1, 1979
Connecticut	- 32	Falls Village	February 16, 1943
Delaware	- 17	Millsboro	January 17, 1893
D.C.	- 15	Washington	February 11, 1899
Florida	- 2	Tallahassee	February 13, 1899
Georgia	- 17	near Calhoun	January 27, 1940
Hawaii	+ 14	Haleakala/Maui	January 2, 1961
Idaho	- 60	Island Park Dam	January 18, 1943
Illinois	- 35	Mt. Carroll	January 22, 1930
Indiana	- 35	Greensburg	February 2, 1951
Iowa	- 47	Washta	January 12, 1912
Kansas	- 40	Lebanon	February 13, 1905
Kentucky	- 34	Cynthiana	January 28, 1963
Louisiana	- 16	Minden	February 13, 1899
Maine	- 48	Van Buren	January 19, 1925
Maryland	- 40	Oakland	January 13, 1912
Massachusetts	- 34	Birch Hill Dam	January 18, 1957
Michigan	- 51	Vanderbilt	February 9, 1934
Minnesota	- 59	Pokegama Dam	February 16, 1903
Mississippi	- 19	near Corinth	January 30, 1966

Missouri	- 40	Warsaw	February 13, 1905
Montana	- 70	Rogers Pass	January 20, 1954
Nebraska	- 47	Camp Clarke	February 12, 1899
Nevada	- 50	San Jacinto	January 8, 1937
New Hampshire	- 46	Pittsburg	January 28, 1925
New Jersey	- 34	River Vale	January 5, 1904
New Mexico	- 50	Gavilan	February 1, 1951
New York	- 52	Old Forge	February 18, 1979
North Carolina	- 29	Mt. Mitchell	January 30, 1966
North Dakota	- 60	Parshall	February 15, 1936
Ohio	- 39	Milligan	February 10, 1899
Oklahoma	- 27	Watts	January 18, 1930
Oregon	- 54	Seneca	January 18, 1930
Pennsylvania	- 42	Smethport	January 5, 1905
Rhode Island	- 23	Kingston	January 11, 1942
South Carolina	- 20	near Longcreek	January 18, 1977
South Dakota	- 58	McIntosh	February 17, 1936
Tennessee	- 32	Mountain City	December 30,1917
Texas	- 23	Seminole	February 8, 1933
Utah	- 50	Strawberry Tunnel	January 5, 1913
Vermont	- 50	Bloomfield	December 30,1933
Virginia	- 29	Monterey	February 10, 1899
Washington	- 48	Mazama & Winthrop	December 30,1968
West Virginia	- 37	Lewisburg	December 30,1917
Wisconsin	- 54	Danbury	January 24, 1922
Wyoming	- 63	Moran	February 9, 1933

The last date of occurrence is printed. Some records also occurred at other locations and/or on earlier dates.

RECORD WORLD EXTREMES

AREA	+°F	LOCATION	DATE
Africa	136	El Azizia, Libya	Sep. 13, 1922**
North America	134	Death Valley, Calif.	July 10, 1913
Asia	129	Tirat Tsvi, Israel	June 21, 1942
Australia	128	Cloncurry, Queensland	Jan. 16, 1889
Europe	122	Seville, Spain	Aug. 4, 1881
South America	120	Rivadavia, Argentina	Dec. 11, 1905
Oceania	108	Tuguegarao, Philippines	Apr. 29, 1912
Antarctica	58	Esperanza, Palmer Peninsula	Oct. 20, 1956

AREA	- °F	LOCATION	DATE
Antarctica	-129	Vostok	July 21, 1983**
Asia	- 90	Oimekon, USSR	Feb. 6, 1933
Greenland	- 87	Northice	Jan. 9, 1954
North America	- 81	Snag, Yukon, Canada	Feb. 3, 1947
Europe	- 67	Ust 'Shchugor, USSR	January +
South America	- 27	Sarmiento, Argentina	June 1, 1907
Africa	- 11	Ifrane, Morocco	Feb. 11, 1935
Australia	- 8	Charlotte Pass, N.S.W.	July 22, 1947*
Oceania	+ 14	Haleakala Summit, Maui, Hawaii	Jan. 2, 1961

+ exact date unknown; lowest in 15-year period

* and earlier date

** world record

Index

ORDER FORM

PHOTOCOPY OR TEAR OUT THIS FORM AND MAIL

CRISIS IN THE ATMOSPHERE:
THE GREENHOUSE FACTOR

Additional copies of this book may be ordered by sending payment for the desired number of copies:

____ # of copies @ $6.95	=	______
Arizona residents add 6.7%	=	______
Shipping & Handling $1.00	=	1.00
Total		______

Send Check or Money Order payable to:

D. B. Clark & Co. Publishing
P.O. Box 35022
Phoenix, AZ 85069

(PRINT) Name ______________________

Address ______________________

City, State, Zip ______________________

Allow 3-4 weeks for delivery
